The Open University

Science: a second level course

ST240

OUR CHEMICAL ENVIRONMENT

BOOK 2
MATERIALS AND ENERGY

The Open University

ST240 Course Team

Course Team Chair	Stuart Bennett	**BBC**	Cameron Balbirnie
General editors	Michael Mortimer		Sandra Budin
	Malcolm Rose		Andrew Law
Authors	Rod Barratt		Paul Manners
	Alan Bassindale		Michael Peet
	Stuart Bennett		Ian Thomas
	Michael Gagan		Nicholas Watson
	Jim Iley		Darren Wycherley
	Michael Mortimer	**Editors**	Gerry Bearman
	David Roberts		Ian Nuttall
	Malcolm Rose		Dick Sharp
	Peter Taylor	**Graphic Design**	Sue Dobson
Course Managers	Simone Pitman		Mark Kesby
	Charlotte Sweeney		Mike Levers
Course Secretaries	Margaret Careford		David Roberts
	Sally Eaton		Howard Taylor
			Rob Williams
		Experimental work	Keith Cavanagh
			Ray Jones
			Pravin Patel

This publication forms part of an Open University Level 2 course. Details of this and other Open University courses can be obtained from the Course Reservations Centre, PO Box 724, The Open University, Milton Keynes MK7 6ZS, United Kingdom: tel. +44 (0)1908 653231, e-mail ces-gen@open.ac.uk.

Alternatively, you may visit the Open University website at http://www.open.ac.uk, where you can learn more about the wide range of courses and packs offered at all levels by the Open University.

To purchase this publication or other components of Open University courses, contact Open University Worldwide Ltd, The Berrill Building, Walton Hall, Milton Keynes MK7 6AA, United Kingdom: tel. +44 (0)1908 858785; fax +44 (0)1908 858787; e-mail ouwenq@open.ac.uk; website http://www.ouw.co.uk

The Open University, Walton Hall, Milton Keynes, MK7 6AA

First published 1995. Reprinted with corrections 2000

Edited, designed and typeset by The Open University.

Printed and bound in the United Kingdom by Bath Press Ltd, Glasgow.

ISBN 0 7492 51425

1.2

ST240book2i1.2

'The Earth is provider of all of beauty and utility'

PART 1 MATERIALS

Prepared for the Course Team by Rod Barratt, Stuart Bennett and Malcolm Rose

Contents

Chapter 5
Polymers

Chapter 1
What are materials?

It is surprising how quickly we accept changes to our lives, not just the landmark changes (which are often of a personal kind) but those changes that creep up on us and yet have a major impact on the way we live. It was only in the 1960s that package holidays really began to take off, and today an overseas holiday is the norm for many people in the UK. This is due in part to the increase in average prosperity but over the past 30 years the real cost of air travel has fallen. In 1964, an economy class return air ticket from London to New York cost £142, representing about *eight times* the average weekly wage at that time. In 1994, the average weekly wage of £260 would buy the ticket for a similar journey.

Figure 1.1
The cost of air travel has fallen dramatically over the past 30 years.

More efficient work practices and economies of scale have helped bring down the cost of flying and there have been major improvements in the design of the planes themselves. New lightweight metal alloys, replacement of metals with plastics, better tyres and engine parts made from new materials are all aspects of the plane that have helped bring down costs (Figure 1.1). These are areas where chemistry and technology play a major role.

The microwave oven has changed cooking habits, and in line with this there has been a notable increase in the range of ready-prepared meals available from food stores. It is not just the application of science and technology of food preparation that has moved forward but the packaging for these meals owes much to the skills of the scientist. The idea of a disposable food container made of plastic that could be used both in a microwave oven and in a conventional oven at a temperature of 200 °C would have seemed fanciful just a few years ago. The transparent film which covers the container has to act as a sterile seal while the pack is on the store's shelf but it must peel off easily even after it has been heated in the oven (Figure 1.2).

Figure 1.2
A ready-prepared meal with packaging that can be used in a microwave oven or heated to 200 °C in a conventional oven.

In 1969, the world watched in awe as Neil Armstrong became the first human to set foot on the Moon. The on-board computers that helped guide the *Apollo 11* mission on its 800 000 km return journey (Figure 1.3) were *less* powerful than the personal computers that sit on desks in almost every office and many homes today. The heart of the modern computer is the microchip based on the element silicon, although new materials such as gallium arsenide are providing the means for even faster-operating machines. Indeed, in the early 1960s a computer of similar power to the desktop of today would have been large enough to fill a small room.

Figure 1.3
'Buzz' Aldrin on the Moon during the 1969 *Apollo 11* mission.

Synthetic fabrics have revolutionized the clothing industry, from the appearance of nylon around 1940 through to the specialist fabrics of today. A commonly used and effective fabric for waterproof clothing is woven nylon coated with a thin layer of another synthetic material, polyurethane. This is completely waterproof and certainly protects the wearer from the rain. However, the body produces a large amount of moisture during periods of exertion. The waterproof nature of this fabric means that this moisture cannot escape and the wearer gets wet with condensed perspiration. What was required was a fabric penetrated with microscopic holes so small that liquid water could not pass through while allowing water vapour to pass from the inside of the garment to the outside. Extensive research produced a polymer material that satisfied these criteria, and this is marketed under the name Gore-tex™. This material has transformed foul weather clothing for hikers and climbers (Figure 1.4), and today there is a range of breathable fabrics based on different polymer materials.

Let's take a moment to look at Figure 1.5, a typical filling station. Make a list of everything that appears in the picture where you think chemistry may have played a part.

Figure 1.4
Breathable yet waterproof outdoor clothing. The secret is in a synthetic layer which allows passage to the exterior of water vapour but excludes liquid water.

Figure 1.5
A petrol station.

The building itself is made of brick, with tiles for the roof, steel for window and door frames and of course glass in the windows. The production of all these materials involves processes that bring about chemical changes. Even the paint used for the various signs on the building has been formulated by chemists. The pumps themselves are made from metal, the displays involve polymers and the fuel hoses are synthetic rubbers. And what about the vehicles? The steel and many other metals and alloys from which they were constructed, the paint and wax used for rust protection, plastics, glass, tyres and lubricants all owe much to the chemist. And, of course, what you cannot see in the picture is the reason for the existence of the petrol station itself – the fuel. Both petrol and diesel fuels are stored in underground tanks and pumped through a metering system into vehicles. These fuels are obtained from crude oil as described in Part 2 of this Book. Even the electricity used to power the pumps is likely to have been generated by burning coal, oil or natural gas in power stations.

Part 1 of this Book has materials as its focus. It aims to show that there is a link between the properties of materials which are useful to us and their structure at the molecular level. One of the great strides of chemistry over the past few decades has been the ability to produce substances with specific properties by making changes at the molecular level. To appreciate how this is done, we need to look at materials in some detail. We discuss the sources of these materials in Chapter 2, while in Chapters 3 to 5 we focus on cement, glass and ceramics, metals, and polymers respectively. The discussion of polymers addresses the technology of polymer production and the problems of recycling as well as the chemistry. The development and use of materials in society are not new. Humans have exploited materials for thousands of years but the pace at which new materials have appeared has accelerated dramatically over the latter part of the 20th century.

Rather than just looking at the nature of developments, it is interesting to speculate on the reasons why particular developments occurred when they did. The impetus that led to the American landing on the Moon was, in part, political. The perceived lead gained in the exploration of space by the former Soviet Union resulted in consternation amongst the political and military leaders of the USA. In a key speech in 1961, President John F. Kennedy promised that a manned landing on the Moon would happen 'within 10 years' and that the astronauts would return safely to Earth. At the time, this seemed to many to be an impossible dream but it was achieved. Politics may have provided the impetus but had a similar statement been made say 20 years earlier, the outcome would have been failure. At that time, the science and technology necessary for success were not available.

Sometimes, it is the need of society for a particular development that provides motivation. Again, the necessary scientific and technological expertise must be available or be developed quickly. In the 1950s, this was the situation that led to a vaccine that gave protection against poliomyelitis, a viral disease of the nervous system. An indication of the effectiveness of the vaccination programme can be seen from Table 1.1, which charts the number of reported polio cases in the UK over a 35-year period from 1955. Once a programme of mass vaccination was started, the incidence of recorded cases fell dramatically. Today, an oral vaccine is used routinely and it has been effective in maintaining a low level of cases.

Table 1.1
Change in the number of reported cases of poliomyelitis in the UK, 1955–90.

Year	No. of cases
1955	6331
1965	91
1975	3
1985	4
1990	4

Society can also benefit through serendipity, a chance discovery or observation that leads to a valuable development. Polymers that can conduct electricity were synthesized in the 1970s and 1980s and are now being used in prototype lightweight batteries for portable computers and electric vehicles. One such conducting material was a type of polymer based on the molecule ethyne which has the formula C_2H_2. During an experiment in Japan, an overestimate of the amount of a reagent in a chemical reaction led to a new silvery material instead of producing the expected white solid plastic. An American scientist from the University of Pennsylvania, Alan MacDiarmid, who was visiting the laboratory at the time, suggested that the electrical conductivity should be measured. It was found to be thousands of times greater than that expected for a plastic and was comparable with that of a metal. As with many breakthroughs, there is usually the need for a shrewd soul to make the link between an observation and a potential application.

But let us go back over a much longer time-span and see what combination of need, knowledge and inspiration has led to changes in the way people have lived throughout history. Take a look at Figures 1.6–1.9 which represent aspects of life in Europe over a span of more than 3 000 years.

Figure 1.6
An artist's impression of Bronze Age metalworking.

Figure 1.7
16th century village life depicted by the Flemish painter Pieter Bruegel the Elder, in his painting of *Netherlandish Proverbs*, 1559.

Figure 1.8 (top right)
London street scene in 1894.

Figure 1.9
A study in a home of the mid-1990s.

The scenes in the pictures are very different: you would expect to see some dramatic changes over a 3 000 year period. However, spend some time looking in detail and try to focus on the changes that have occurred over this period. Analyse each picture under the following headings: materials for building, tools used, transport, clothing, colour, and food. (You will not be able to extract information under all the headings from each picture.)

The picture of the Bronze Age village in Figure 1.6 is based on archaeological evidence for around 1400 BC. Information that can be gleaned from the picture is limited but it is clear that the materials used in the construction of the dwellings are not highly processed. Roofs are thatched with grasses and reeds on a crude wooden frame. Wood, stone and earth are used for building walls. Clothing is roughly woven plant material or animal skin and food probably comes from small cultivated plots around the village. There is evidence of small-scale processing of materials from the bronze tools and fired clay cooking pots. Needless to say, colour and transport do not feature strongly, with the inhabitants striving to maintain what seems to be a somewhat drab life in the environs of the settlement.

A gap of about 3 000 years separates our Bronze Age village from the picture of a Flemish village in Figure 1.7, painted by Pieter Bruegel the Elder in 1559. This picture gives a good indication of village life in northern Europe in the 16th century but it must be regarded as something of a caricature. It might seem incredible but over 100 proverbs, sayings and customs of the time have been identified in the picture although only a handful will be familiar to contemporary British culture. For example, 'banging one's head against a brick wall' or 'filling a hole after a calf has drowned' (equivalent to shutting the stable door after the horse has bolted) may strike a chord but it is unlikely that 'a man shears a pig and another a sheep' is in current parlance.

Our immediate interest is not with these maxims but with the evidence that the picture gives us for the way materials were used in the 16th century. Look first at the building materials: wood, stone, brick, tile, glass and lead in the windows. Many of the tools are metal. The shovel and shears are probably iron or steel and it appears that there are metal hinges on window shutters. Transport is depicted by a sailing boat constructed of wood with a sail made of woven cloth, probably a canvas of flax from the fibre of the linseed plant. There is also a wooden cart wheel with a metal rim. Villagers are dressed in a range of woven clothing, some of which may be wool from domesticated sheep. A bleach must have been used to achieve the whiteness of shirts and headgear, and dyes have given a range of colours to other clothing. The colours are soft and muted which suggests that the dyes may have been extracted from plant materials. However, this is a picture that was painted over 400 years ago when the availability of painting pigments was limited. It may be, too, that the colours of these paints have faded with time. Food is represented by the crops, possibly wheat, shown growing in a field and from the sheep, pigs and the calf. The scene is of a relatively self-sufficient village but with longer distance travel and communication afforded by the sailing ship.

Another 300 years or so and we are in the latter part of the Industrial Revolution (Figure 1.8). Buildings are of brick, stone and tiles, not so different from today. The range of clothing available to people has moved on since the 16th century and there is evidence of the use of metals in the lamp post and the carriage. At this time, many people worked in factories and mills where there were machines, looms and carding machines of the cotton industry. The age of steam had arrived with railways and steam-powered machinery in factories. During the Industrial Revolution there was a movement of population from the countryside into towns and cities.

And then we move on to today (Figure 1.9) where there are manufactured materials wherever we care to look – metals, plastics and ceramics. The two centuries since the onset of the Industrial Revolution have seen a huge number of changes, far more than were achieved over the previous 3 000 years. The real difference is that now we can develop materials with almost whatever properties we require.

We begin our story by exploring how the Earth provides the potential to produce the materials for modern life.

Chapter 2
Where do materials come from?

In all four pictures (Figures 1.6–1.9), metals and stone are featured. This reflects perhaps two characteristics of humans: the use of tools and the need for shelter. During the Bronze Age, stone for building was simply taken from the Earth and roughly shaped, whereas the building bricks of the Victorian era involved processing of clay. Similarly, the only metals available in the Bronze Age were gold, silver and bronze (actually a mixture of two metals, copper and tin). In the 16th century, the major metal used was iron or steel but as we get closer to today that range of metals has increased greatly. Most of the materials that are in use today, and that have been used over history, have been processed. This processing may be crude, as in shaping stone for simple buildings, or, as you will see later, it might be the sophisticated techniques that are needed for modern materials.

Coal, water, iron, stone and clay (for bricks) are all familiar materials that have shaped society in Britain, particularly over the past 200 years. These resources were relatively abundant in Britain because of both geology and climate. The way that resources are distributed on Earth helps us to understand how materials have influenced history and how humans can use resources to design materials for a better future.

2.1 Elements of the Earth

A widely accepted view is that the Universe arose from the Big Bang about 15–20 billion years ago. The first two elements to be formed in the aftermath of the explosion were hydrogen and helium and these gases eventually condensed under gravitational forces to form galaxies and the stars within them. The extraordinarily high temperatures inside stars resulted in the formation of many of the elements that we know today. Even so, much of the matter in the Universe is in the interstellar gas clouds of hydrogen and helium, and that is reflected in the **distribution of the elements** shown in Figure 2.1a. The Earth was formed from condensed material around the Sun and contains a range of elements.

The Earth has a diameter of about 12 700 km but the crust is relatively thin, being between 25 and 70 km deep on land. (The thicker parts tend to be under the major mountain ranges.) Under the

Figure 2.1
Distribution of elements by mass in (a) the Universe, (b) the Earth overall and (c) the Earth's crust. (Note that these figures can vary according to the source of the data and measurement methods.)

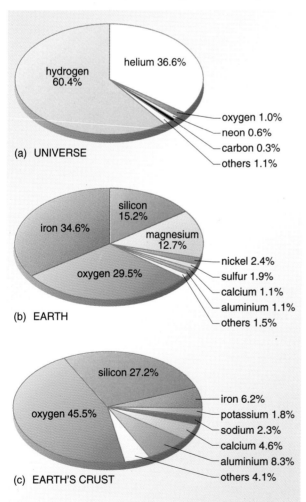

(a) UNIVERSE

hydrogen 60.4%
helium 36.6%
oxygen 1.0%
neon 0.6%
carbon 0.3%
others 1.1%

(b) EARTH

iron 34.6%
silicon 15.2%
magnesium 12.7%
oxygen 29.5%
nickel 2.4%
sulfur 1.9%
calcium 1.1%
aluminium 1.1%
others 1.5%

(c) EARTH'S CRUST

silicon 27.2%
oxygen 45.5%
iron 6.2%
potassium 1.8%
sodium 2.3%
calcium 4.6%
aluminium 8.3%
others 4.1%

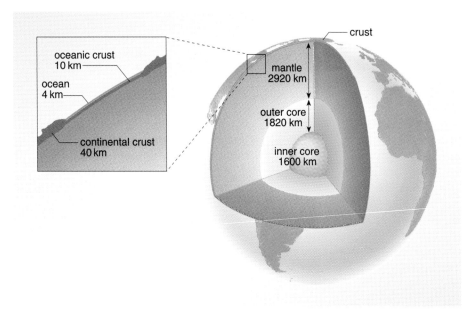

Figure 2.2
The structure of the Earth. The Earth's crust is a thin skin on the surface of the planet, while the bulk of the Earth comprises a stony mantle enclosing a central core. The inner core is solid but the outer core is thought to be liquid.

oceans, the crust is between 8 and 12 km thick. Even with modern technology, we can only extract solid materials from the upper part of the crust. Figure 2.2 indicates just how much of the Earth is inaccessible to direct exploration. The deepest extraction is for oil, and commercial wells have been drilled to depths of up to 7 km.

▨   Look again at Figure 2.1. What is the first thing that you notice about the distribution of elements in the Universe and in the Earth overall?

■   An obvious point is that the distributions are different. For example, the element hydrogen comprises 60.4% of the mass of the Universe yet in the Earth overall it is iron that is the most common element at 34.6%.

The various elements are not evenly distributed in the Earth however, as you can see from Figure 2.1. Compare the relative proportions of the elements in the whole Earth (Figure 2.1b) and in just the outer layer, the Earth's crust (Figure 2.1c).

▨   What element shows the greatest difference in proportion in the whole Earth and the Earth's crust distributions?

■   Iron. In the whole Earth, almost 35% of the mass is iron yet in the crust the proportion is about 6%.

In its very early history, when the Earth was much hotter and molten, the more dense elements like iron tended to sink towards the centre of the Earth under the influence of gravity. Today, the Earth has a core comprised largely of iron with some nickel, while the solid crust is relatively depleted in iron. There is another difference: iron at the core of the Earth is elemental – it is not chemically combined with any other element. In contrast, iron near the surface of the Earth has had the opportunity to react with oxygen to form iron oxides, compounds of iron and oxygen. Iron is found near the Earth's surface as the metal oxide and not as the element.

Box 2.1 Representing data

The data in Figure 2.1 are displayed by what are commonly known as **pie charts**. The area of the segment represents the same proportion of the whole 'pie' as the quantity of the featured element represents of the total. So, in Figure 2.1c, which represents the distribution of elements in the Earth's crust, the element silicon shows as just over one-quarter of the 'pie' (marked as 27.2%). About one-quarter of the mass of all the atoms in the Earth's crust is accounted for by silicon atoms.

There are other ways of representing data, with which you may be familiar. One of these is the table or tabular representation which you met in Book 1. A table has a number of headed columns.

■ Try to construct a table that shows the data in Figure 2.1c. List the elements in alphabetical order by name but note that it is usual to place 'others' last.

■ An appropriate table would have two columns headed 'Element' and 'Proportion by mass', like Table 2.1.

Table 2.1
Distribution of elements in the Earth's crust by mass (in alphabetical order by name).

Element	Proportion by mass
aluminium	8.3%
calcium	4.6%
iron	6.2%
oxygen	45.5%
potassium	1.8%
silicon	27.2%
sodium	2.3%
others	4.1%

Listing the elements in Table 2.1 in alphabetical order of names can be helpful if you wish to quickly locate an element in a long table, but does it give you a good picture of the distribution of the elements?

■ Which representation do you feel makes it easier to identify the most abundant element – the pie chart or the table?

■ The large proportion of oxygen is easy to pick out from the pie chart.

However, Table 2.1 could have been compiled so that the elements were arranged according to their distributions. This is done in Table 2.2, which has the most abundant element, oxygen, at the top of the table, but note that 'others' is still generally placed last.

Table 2.2
Distribution of elements in the Earth's crust by mass (in descending order of abundance).

Element	Proportion by mass
oxygen	45.5%
silicon	27.2%
aluminium	8.3%
iron	6.2%
calcium	4.6%
sodium	2.3%
potassium	1.8%
others	4.1%

■ Does this format have any advantages or disadvantages?

■ It is now much easier to identify the most and least common elements. . But note that to locate a particular element is not so convenient as it is in the alphabetical order list.

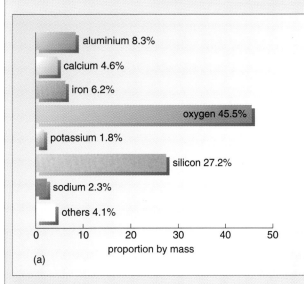
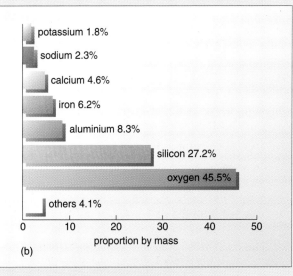

Figure 2.3
Histogram showing the distribution of elements in the Earth's crust by mass in (a) alphabetical order by name and (b) increasing order of abundance.

Another common way of displaying data is in the form of a **histogram**, as shown in Figure 2.3. Histogram representations do make it very clear which elements have the highest or lowest abundance, and individual elements are easy to pick out if alphabetical order is maintained. The choice of how data are represented depends on just what is required to be emphasized and the way in which the data are to be used. You will meet these methods and others as you progress through this Course.

There is, however, a cautionary note, the significance of which will become more apparent later. Figure 2.1 shows the distribution of elements in the Earth's crust and it indicates that the distribution data are based on measurements of mass. Volume could equally well have been chosen as a basis for these measurements but would it have made any difference?

Iron is a rather dense metal with 1 cm³ of iron having a mass of about 7.9 g. Aluminium is much less dense having a density very close to 2.7 g per cm³ which we can represent as 2.7 g cm⁻³.

You may recall from Book 1 that 10^{-3} is equivalent to $\frac{1}{10^3}$; this convention can also be applied to units. So for aluminium, 2.7 g per cm³, $\frac{2.7\,g}{cm^3}$ and 2.7 g cm⁻³ are all equivalent ways of expressing the density.

Supposing that we had a container in which there was 1.0 cm³ of iron and 2.0 cm³ of aluminium.

▨   If you were asked in which container was there more metal, how would you respond?

▨   The answer must be that 'it depends'. If volume is the criterion, then there is more aluminium. The volume of the aluminium is twice that of the iron. But what about mass? The mass of the 2.0 cm³ of aluminium is about 5.4 g. The 1.0 cm³ of iron has a mass of 7.9 g. So using a mass criterion, there is more iron.

The distributions of Figure 2.1 are based on mass and you will have noticed that this is specified in the figure heading.

Gold is unusual among the metals in that it exists in the Earth's crust as the element and not chemically combined with other elements. You saw in Book 1 that gold is resistant to chemical attack by air and water. (The only other metals which are sometimes found as the element are silver and copper.) Gold is also a relatively rare element. An entry for gold in Table 2.1 would indicate an abundance of only 0.000 000 4% by mass.

▨   Write the abundance of gold in the Earth's crust in scientific notation.

◼   The abundance is 4×10^{-7} % by mass.

Another way of representing this figure, and sometimes a more convenient way, is to quote the number of parts of gold in a number of parts of the Earth's crust. We are using mass as our criterion so what we need to know is 'how many kilograms of the Earth's crust contain one kilogram of gold?'.

▨   What fraction of the Earth's crust is gold?

◼   Proportion by mass as a percentage is equal to the mass of the element (in kg) that is in 100 kg of the crust. If an element were 2% abundant then there would be 2 kg of the element in 100 kg of the Earth's crust. The fraction of the element in the crust would therefore be $\frac{2}{100}$.

The fraction of gold in the Earth's crust is then

$$(4 \times 10^{-7}) \times \frac{1}{100} = (4 \times 10^{-7}) \times 10^{-2} = 4 \times 10^{-9}$$

This can be written as $\dfrac{4}{1 \times 10^9}$

There are four parts of the element gold in every 10^9 parts of the Earth's crust by mass, i.e. there would be $4\,kg$ of gold in $10^9\,kg$ of the Earth's crust.

Question 1 If all the elements in the Earth's crust were evenly distributed, how much earth would be needed to extract $20\,g$ of gold? (This mass is approximately represented by a cube of side 1 cm.)

Some idea of what the calculation in the answer to Question 1 implies can be seen from Figure 2.4.

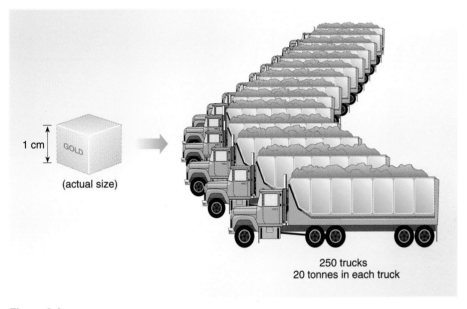

Figure 2.4
If gold were evenly distributed through the Earth's crust, the amount of earth that would be needed to extract $20\,g$ of the element would fill 250 twenty-tonne trucks.

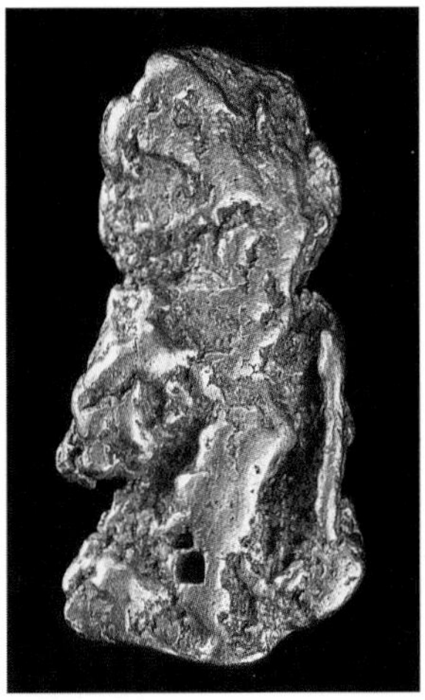

If you have had difficulty with this and the preceding calculation, you should listen to AV sequence 'Simple maths' at this point.

Gold is not, however, distributed evenly throughout the crust of the Earth and neither are the other elements. The 19th century gold rushes of California (1849) and the Klondike in Western Canada (1897) would certainly never have happened had there been an even distribution of the elements. In both areas, the rush for gold started with the chance finding of nuggets weighing several grams (Figure 2.5). It would have been out of the question for an individual prospector to shift 5 000 tonnes of earth even if the technology were available to extract the $20\,g$ of gold.

▨ What are the two most common elements (by mass) in the Earth's crust?

■ From Figure 2.1c (or Table 2.1 or 2.2), oxygen and silicon are the most common elements. Oxygen is 45.5% and silicon is 27.2% abundant by mass.

Figure 2.5
A gold nugget.

Silicon is a lustrous, metallic grey solid (Figure 2.6) and oxygen, a colourless gas.

Figure 2.6
Elemental silicon.

Again it is important to know that we are using comparisons of mass and not volume. The mass of a cube of silicon of side 1.0 cm is about 2.3 g but the same volume of the gaseous element oxygen would have a mass of just over 0.00 3 g under normal temperatures and pressures. Our figures indicate that the *mass* of oxygen in the Earth's crust is almost twice that of silicon. The reason for the large quantities of silicon and oxygen in the Earth's crust is that many rocks are composed of these elements. Common sand as found on so many beaches and in deserts (Figure 2.7) is a compound of silicon and oxygen known as an oxide of silicon.

Figure 2.7
Sand dunes in California, USA.

■ What can you deduce about the physical characteristics of a compound by the properties of the elements from which it is formed?

■ Taking the case of sand, the answer must be *nothing*. Sand looks nothing like the grey solid element, silicon (Figure 2.6), or the colourless gas, oxygen. (However, armed with the ideas of chemistry that you will develop in this Course, you will be able to make many useful predictions.)

Box 2.2 What is white?

Sand would look white if it were not for the impurities of iron (mainly an iron oxide) that give the characteristic pale yellow–brown colour. However, a chemist would describe the pure oxide of silicon as colourless and not white. There is certainly a difference between colourless and white to our eyes so how does this apparent confusion arise?

Water is essentially colourless and so is window glass (which is mainly an oxide of silicon). These materials are also transparent in that it is possible to see images through them with little distortion. Now, what would a piece of window glass look like if it were pounded into tiny fragments with a hammer? The outcome is a white powder that is no longer transparent. Its colour properties have not changed. Under a magnifying lens, each individual particle of glass would still appear to be transparent and colourless yet collectively the particles appear white (Figure 2.8).

You do not have to smash glass to see this effect. Examine some grains of table salt under a magnifying lens. Although salt looks to be

(a)

(b)

Figure 2.8
A piece of window glass (a) before and (b) after pounding with a hammer. The unbroken glass appears transparent and colourless but the powdered glass is white.

white in bulk, the individual grains appear transparent and colourless.

Light hitting the fragments of glass is reflected from the surfaces. Some light may pass through a few fragments but will soon hit a surface at an angle that will result in reflection. Multiple reflections of light between the fragments are responsible for the white appearance (Figure 2.9).

White paint appears white for a somewhat different reason. Much white paint contains particles of an oxide of the metal titanium. These oxide particles are transparent and have a marked ability to bend light. (This is the same effect that can be observed when a flower stem stands in a glass vase of water.) The path of light changes as it passes through the particles and comes back from the painted surface (Figure 2.10). This process is known as refraction but the result is a white appearance like the reflective process in Figure 2.9.

Colour, how it arises and how dyes are made is a major part of Book 4.

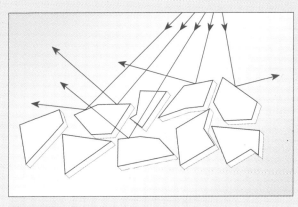

Figure 2.9
Multiple reflections of light in powdered glass.

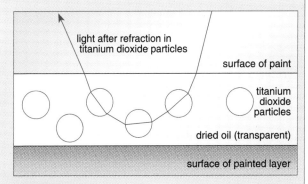

light after refraction in titanium dioxide particles

surface of paint

titanium dioxide particles

dried oil (transparent)

surface of painted layer

Figure 2.10
Light refracted by particles of titanium dioxide in paint to give a white surface.

2.2 Minerals

All rocks are made of minerals. Minerals are naturally occurring chemical compounds of fixed composition. The most common minerals are those containing silicon and oxygen. In Book 1, the Periodic Table shows silicon to be in the same group as carbon. Like carbon, silicon tends to form four bonds when it combines with other elements. In the pure element, each silicon atom is linked to four similar atoms through single chemical bonds, as indicated in Figure 2.11.

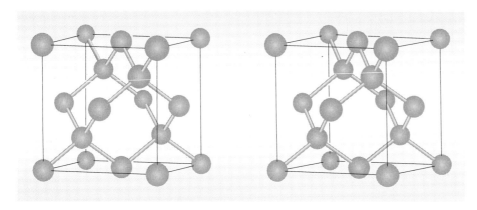

Figure 2.11
Linking of atoms in elemental silicon.

Oxygen exists as a gas comprising molecules in which two oxygen atoms are linked by two bonds (Figure 2.12). Based just on the idea of silicon forming four bonds and oxygen two bonds, there are two main ways in which molecules containing silicon and oxygen can be assembled. One possible structure is shown in Figure 2.13. It is a similar structure to that of carbon dioxide with the silicon atom being linked to two oxygen atoms through double bonds. In fact, it is found that this particular molecule does *not* exist. This is because silicon, unlike carbon, is reluctant to form double bonds and prefers a structure like that shown in Figure 2.14 in which silicon is linked to oxygen atoms through single bonds.

Figure 2.12
The oxygen molecule, showing the double bond linking the two oxygen atoms.

Figure 2.13
A possible structure in which a silicon atom is bonded to two oxygen atoms by double bonds.

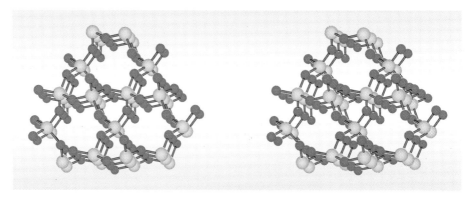

Figure 2.14
Stereostructures showing silicon and oxygen atoms joined by single bonds. Silicon is in grey and oxygen is in red.

To maintain four bonds to silicon and two to oxygen, each of the two structures in Figures 2.13 and 2.14 necessarily has twice as many oxygen atoms as silicon atoms. The actual structure adopted by the compound that we can now call silicon *di*oxide is that of Figure 2.14. (The *di* in silicon *di*oxide indicates that there are *two* oxygen atoms for every silicon atom. It is possible to construct a molecule containing a mixture of single and double bonds. However, such structures have not been found to exist in naturally occurring silicon–oxygen compounds.)

The mineral quartz is pure silicon dioxide. Quartz forms colourless transparent crystals of characteristic shape (Figure 2.15); its crystal structure is shown in Figure 2.16.

Figure 2.15
Single crystal of quartz

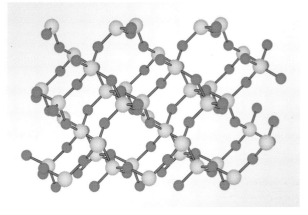

Figure 2.16
Section of the structure of quartz. Oxygen atoms are shown in red and silicon atoms in grey.

Quartz is sometimes found with small quantities of metals as impurities and these metals often impart beautiful colours. The distinctive colour of amethyst (Figure 2.17) arises from tiny quantities of iron as an impurity. Agate too (Figure 2.18) is an impure form of silicon dioxide; and flint is mainly silica.

Figure 2.17
Crystals of amethyst.

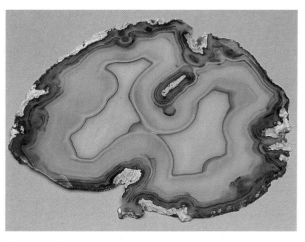

Figure 2.18
Agate, showing its characteristic bands of colours.

Box 2.3 Chemical formulas

You will realize by now that **chemical formulas** are part of the language of chemistry and can appear to be rather complex. A little effort in coming to terms with formulas can open the door to so much chemistry. We shall extend our ideas further at a number of stages later in this and subsequent Books.

The word 'formula' has a Latin origin meaning 'form' or 'shape'. The plural is formulae if one adheres to the Latin origin and this is much used in scientific writing. However, we shall use the plural form which is becoming increasingly common, formulas.

A piece of quartz of mass 10 g (about enough to fill a teaspoon) contains around 1×10^{23} atoms of silicon and twice that number of atoms of oxygen.
A representation of this mass of 10 g quartz on the scale of that in Figure 2.16 would be a cube with side of roughly 750 km, the

distance from London to Zurich in Switzerland. So, we necessarily have to represent only a portion of the structure. This is justified, as structures of pure crystalline compounds are repeating patterns of small units known as **unit cells**, an idea you first met in Book 1. But even the limited ball-and-stick model takes up space on the page and, in many instances, it is sufficient to know the relative numbers of the atoms.

▓   Count the number of silicon and oxygen atoms in Figure 2.16, which shows a representative section of the structure of quartz.

▓   You should have counted 39 silicon and 78 oxygen atoms.

In a chemical formula, the numbers of the atoms are represented by subscripts written immediately after the symbol for that particular atom. So, the structure in Figure 2.16 could be

represented by the formula $Si_{39}O_{78}$. Can you see a problem with this formula?

The formula $Si_{39}O_{78}$ shows that there are twice as many oxygen atoms as there are silicon atoms. The numerical subscripts in the formula seem to depend on just how much of the extended structure is shown in the Figure. Another author may have used a different representation with perhaps 50 silicon atoms. The formula then would be written as $Si_{50}O_{100}$.

It seems an unnecessary complication to have different formulas for the same compound and this is generally to be avoided. The formula that is used is the **empirical formula** which reports the relative numbers of atoms of each element present in the compound. The empirical formula for quartz is SiO_2, and this is reflected in the name silicon dioxide. (Where there should be a subscript 1, it is usually omitted so that the formula is written SiO_2 and not Si_1O_2.)

The whole structure of a single quartz crystal can be represented as a giant molecule with millions of silicon and oxygen atoms held together with chemical bonds – it has an **extended structure**. This is in contrast to discrete molecules of carbon dioxide which each contain one carbon and two oxygen atoms. The empirical formula for carbon dioxide is CO_2 and the **molecular formula**, which indicates the *actual* numbers of carbon and oxygen atoms in a carbon dioxide molecule, is also CO_2. The empirical formula for quartz is SiO_2 but the molecular formula is not defined.
You should now listen to AV sequence 'Talking chemistry' in which we discuss the ways that the chemist uses formula representation.

Question 2   What are the empirical formulas of the following: C_8H_{16}, Zn_4S_4, C_2H_6, Ca_4F_8 and $Cu_4Fe_4S_8$?

The naming of substances with the empirical formula SiO_2 is a little confusing. The chemical name is silicon dioxide, a name that reflects the empirical formula. Unfortunately, the name silica is also used for substances of empirical formula SiO_2. Keep in mind that the names silicon dioxide and silica are equivalent. Silicon is the name of the element. The situation is further complicated by the fact that silica is found in nature in a number of different forms. Quartz is one such form and the minerals cristobalite and lechatelierite are others. All of these minerals are extended structures in which silicon atoms are linked to oxygen atoms through single chemical bonds – they are all silicon dioxide. The different crystal structures arise from detailed differences in the way that the atoms are arranged relative to one another.

Figure 2.19
Mica, showing its layer structure.

Figure 2.20
Two very different forms of carbon: (top) diamond, showing both rough and polished stones; and (bottom) graphite.

Figure 2.21
The molecular structure of (a) diamond and (b) graphite. Carbon atoms are represented by the black spheres.

A quartz crystal will shatter into small particles if it receives a sharp blow with a hammer. Mica is a mineral based on a silicon–oxygen–aluminium structure. A characteristic of the structure is that it forms sheets rather than the three-dimensional array as we have in quartz. This molecular structure is reflected in the piece of mica in Figure 2.19. The layers can be seen clearly and mica can be split to give thin sheets of the mineral. These sheets are transparent and are used as windows for small furnaces. Talc has a similar layer structure to that of mica but it does not contain aluminium. Its slippery feel comes about because these layers are held together only by weak forces and are able to slide easily over each other. The variety of ways in which the silicon and oxygen can be linked together, particularly when other elements are incorporated into the structure, gives a huge range of materials which include many rocks.

You can now begin to see the tremendous variety of materials that can be formed from just a few different elements. Taking all the elements and the myriad of ways in which they can be linked together, it is little wonder that there are over ten million known chemical compounds, and almost certainly many more that have yet to be identified and synthesized.

Not all possible structures for different molecules are realized. The conditions of temperature and pressure often have a significant influence. Even some of the elements have different structures. You have seen that carbon is be found in the Earth's crust in a number of guises. Diamond and graphite are pure forms of carbon (Figure 2.20). Diamond occurs where carbon has been subject to temperatures of over 3 000 °C and pressures above 100 000 bar. Where conditions have been much less severe, graphite is the norm.

Question 3   The structures of two forms of carbon – diamond and graphite – are shown in Figure 2.21. What can you recall about their physical properties and can you relate this to the structures?

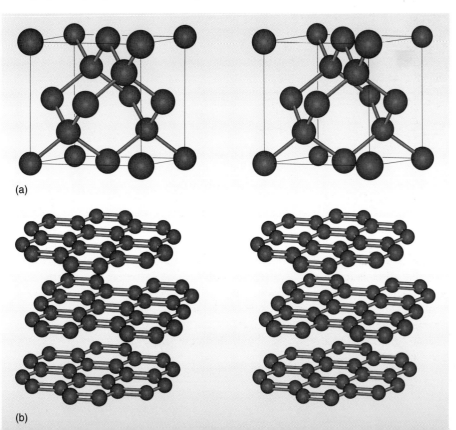

(a)

(b)

Chemists have a problem in representing extended structures on the page. Extended structures present two challenges: how much of the extended structure it is necessary to show and how the structure can be represented in three dimensions.

The mineral chalcopyrite (Figure 2.22) is a compound containing iron, copper and sulfur and it has been used throughout history as a source of copper. Chalcopyrite has an extended crystalline structure. The rather complicated unit cell is cuboid (an elongated cube) and is shown in Figure 2.23.

Figure 2.22
Chalcopyrite.

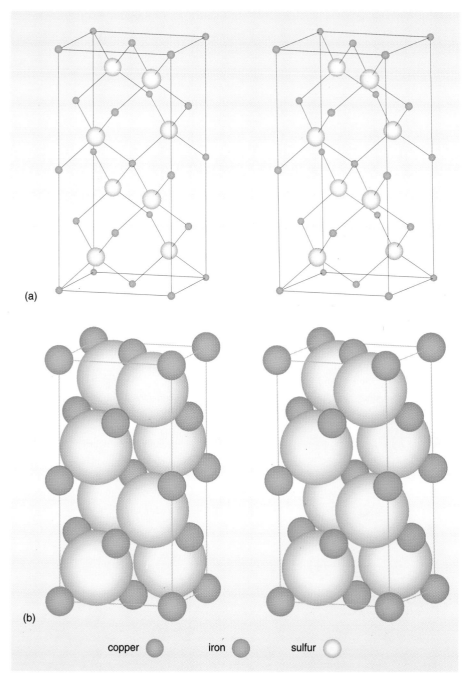

(a)

(b)

copper ●   iron ●   sulfur ○

Figure 2.23
Stereo structures of the unit cell of chalcopyrite: (a) shows the location of the atoms and (b) indicates the relative sizes of the atoms. Note that the lines connecting atoms do not represent chemical bonds but are there to indicate the positions of the atoms in space.

Unit cells may contain atoms or ions. In the case of sodium chloride, which was introduced in Book 1, it was made clear that the unit cell comprised sodium and chloride ions. In some situations, the distinction between atoms and ions in a solid structure is not so clear cut. In general, we shall use the term atom and only use ion where the distinction is unequivocal.

▨ How many copper, iron and sulfur atoms are represented in the unit cell?

▨ Figure 2.23 shows 13 copper atoms, 10 iron atoms and 8 sulfur atoms.

The empirical formula of chalcopyrite is $CuFeS_2$ and not $Cu_{13}Fe_{10}S_8$ as it might appear at a first glance at Figure 2.23. You may have guessed from the discussion of unit cells in Book 1 that this is because the atoms at the boundaries of the unit cell are shared with adjacent unit cells. For example, in chalcopyrite, a copper atom at the corner of the unit cell is not just part of the unit cell in Figure 2.23 but is shared with the corners of adjacent unit cells. Perhaps the best way to see this is to use a set of cubes. Sugar lumps are ideal.

You will need eight sugar cubes. Mark one corner of each cube with a felt-tip pen. Next, take four cubes and arrange them so that they are form a square with the marked corners touching (Figure 2.24).

The marked corners indicate that the atom is shared by these adjacent cubes. But this is not the complete picture, for there is another layer of cubes on top of this one in our representation of an extended structure. Then put together the remaining four cubes just as was done for Figure 2.24 with the marked corners touching. Place this group of four cubes on top of the first four so that so that all eight corners touch, as indicated in Figure 2.25. There is just the one 'atom' forming the corner of eight cubes. Thus, an atom at the corner of a cuboid unit cell is only contributing $\frac{1}{8}$ to that unit cell. The rest of the volume of the atom, $\frac{7}{8}$, is divided among the remaining seven unit cells that touch at that point.

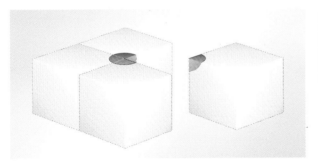

Figure 2.24
Four sugar cubes representing unit cells.

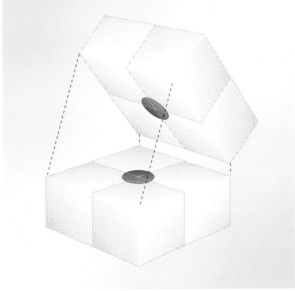

Figure 2.25 (right)
Assembly of eight sugar cubes.

▓   Look again at Figure 2.23. Are there atoms other than those at the corners that are shared with adjacent unit cells?

■   There are two additional shared environments. There are iron atoms on the edges of the unit cell half-way down. There are also both iron and copper atoms on the faces of the unit cell.

▓   Use the sugar cubes to determine the number of unit cells that share an edge atom and the number that share a face atom.

■   You should have found that an edge atom is shared by four cells and a face atom by two cells.

We are now in a position to again recount the relative numbers of atoms in the unit cell of chalcopyrite using our system of counting atoms shared with adjacent unit cells.

Remember,

●   an atom at the corner of a unit cell counts $\frac{1}{8}$;

●   an atom on the edge of a unit cell counts $\frac{1}{4}$;

●   an atom on a face of a unit cell counts $\frac{1}{2}$;

●   and, of course, an atom wholly within a unit cell counts 1.

To do our counting, we need to study Figure 2.26 which shows the unit cell of chalcopyrite with the different atom environments highlighted.

We shall look at each atom in turn, starting with the simplest environment, that of sulfur. The unit cell contains eight sulfur atoms which are wholly within the boundaries of the unit cell. These atoms are not shared with any adjacent unit cells and make a contribution of eight sulfur atoms.

Iron atoms appear in two distinct environments, on the edges and on the faces of the unit cell. There are four iron atoms on the edges each counting $\frac{1}{4}$, and six atoms in face positions each counting $\frac{1}{2}$. So for iron there are $(4 \times \frac{1}{4})$ plus $(6 \times \frac{1}{2})$, a total of four iron atoms.

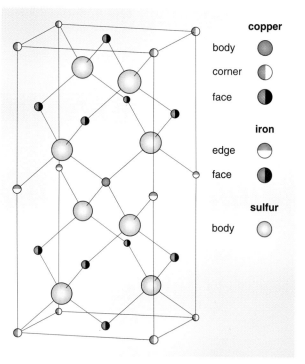

Figure 2.26
Atom environments in the unit cell of chalcopyrite.

Copper atoms appear in three environments: corner, face and in the body of the unit cell. There are eight corner copper atoms each counting $\frac{1}{8}$ and four face copper atoms each counting $\frac{1}{2}$. The copper atom wholly contained within the boundary of the unit cell makes a contribution of one. For copper, then, there are $(8 \times \frac{1}{8})$ plus $(4 \times \frac{1}{2})$ plus 1 making a total of four copper atoms.

Overall, the atoms in a unit cell of chalcopyrite can be represented by the formula $Cu_4Fe_4S_8$, which we define as the **unit cell formula**.

■ What is the empirical formula of chalcopyrite?

■ The empirical formula is the simplest formula that indicates the relative numbers of atoms of each element in the substance. The empirical formula for chalcopyrite can be obtained by dividing the unit cell formula by four to give $CuFeS_2$.

It is important to note that the unit cell formula for a crystalline compound is always a multiple of the empirical formula. The unit cell formula of chalcopyrite $Cu_4Fe_4S_8$ is related to the empirical formula $CuFeS_2$ by a factor of four.

Figure 2.27
Unit cell representations of
(a) caesium chloride,
(b) zinc sulfide and
(c) calcium fluoride.

Question 4 What are the unit cell formulas and the empirical formulas of the compounds represented by the unit cells in Figure 2.27?

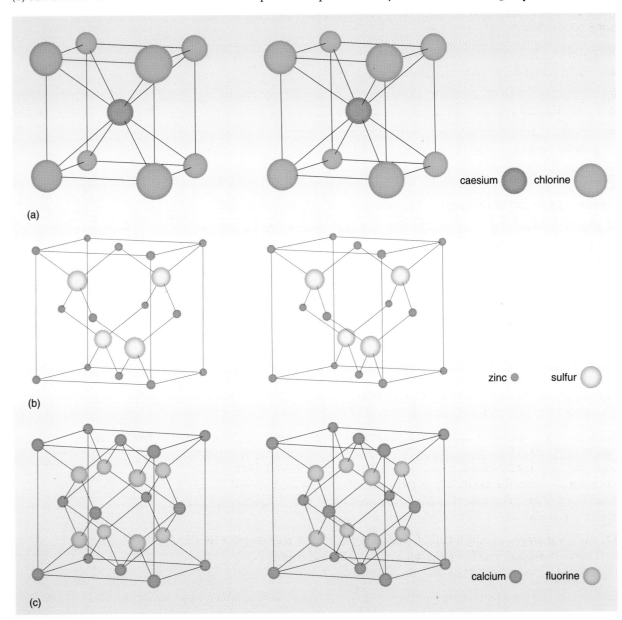

(a)   caesium ●   chlorine ○

(b)   zinc ●   sulfur ○

(c)   calcium ●   fluorine ○

For all minerals, the question that must be asked is why should the minerals have separated and not be spread evenly through the rocks. We have already seen that the low abundance of gold (4×10^{-7}% by mass) would make the metal virtually impossible to obtain if it were distributed evenly through the Earth's crust. Many minerals are found in a relatively pure form. Chalcopyrite was mined extensively in England from about 1600 until the early part of the 20th century. It is found in narrow veins in rocks from parts of the Lake District and Peak District of England (Figure 2.28).

In the early stages of its development, the rocks of the Earth were molten. A whole range of elements and compounds was mixed in with the molten rocks or magma as it is known. In time, the Earth cooled and the outer surface began to solidify. It is at this stage that the separation of different minerals began and this is a continuing process, particularly in areas of volcanic activity.

Figure 2.28
Narrow veins of chalcopyrite.

Magma is a complex mixture of materials including metals, compounds of metals with oxygen and with sulfur and a range of silicon–oxygen compounds all above their melting temperatures at say 3 000 °C. We can take a simplified representation of a magma which is about 90% by mass silicon dioxide SiO_2, 5% aluminium oxide Al_2O_3 and 5% chromium oxide Cr_2O_3. Melting temperature and density data of these oxides are indicated in Table 2.3.

Table 2.3 Normal melting temperature and density data for some oxides.

Compound	Normal melting temperature/°C	Density/g cm^{-3}
Cr_2O_3	2 435	5.2
Al_2O_3	2 015	4.0
SiO_2	1 700	2.3

Box 2.4 Units in tables

Take a close look at Table 2.3 and, in particular, at the column headings for normal melting temperature and density. The second column of the table is headed 'Normal melting temperature/°C' and the entries in the column below the heading are simply numbers.

The normal melting temperature of chromium oxide (Cr_2O_3) is 2 435 °C. However, it would become rather tedious if we had to put the temperature units (°C) after each entry in this column. To avoid this, both the column heading and the quantity to be entered in the column are divided by the units. It is rather like the way that one can treat a mathematical equation by doing to one side what one does to the other.

The equation is
normal melting temperature of Cr_2O_3 = 2 435 °C

Dividing both sides of the equation by '°C' gives

$$\frac{\text{normal melting temperature of } Cr_2O_3}{\text{°C}} = \frac{2\,435\,°C}{°C} = 2\,435$$

Rather than writing fractions in the column heading, it is more convenient to use the solidus or slash '/' which has exactly the same meaning as does the horizontal line in a fraction. Writing the column heading in this way avoids the need to repeat the units for every entry in the column.

The third column is headed 'Density/g cm^{-3}' and the entry for Cr_2O_3 is 5.2. We can effectively read an equation from the table:
density/g cm^{-3} = 5.2
If we multiply both sides of the equation by the unit, g cm^{-3}, we have the simple relationship:
density = 5.2 g cm^{-3}

- What do you expect would happen as our simplified magma cools?

- When the temperature drops to 2 435 °C, chromium oxide will begin to solidify and separate from the liquid. As the temperature continues to fall, aluminium oxide will separate at around 2 015 °C with the bulk of our magma remaining liquid.

The differing densities of the oxides implies that gravity will lend a helping hand in separating oxides as they solidify.

Once the temperature of our magma drops below about 2 000 °C, the two metal oxides will have solidified and the remaining magma, largely silicon dioxide, will remain as a liquid. The greater density of aluminium and chromium oxides helps the solids sink through the liquid magma and collect as a metal-rich deposit (Figure 2.29).

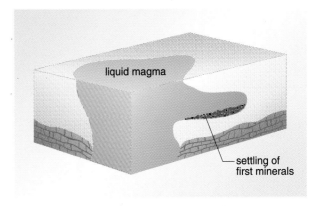

Another method by which minerals separate from magma involves water. Ground water may be heated as it comes into contact with cooling rock. This hot water is more effective at dissolving some minerals from the rock than is cold water. As it flows to cooler areas, these minerals come out of solution and are selectively deposited (Figure 2.30). Sometimes it may be that magma under pressure is forced into cracks in surrounding rocks. Slow cooling gives rise to veins of mineral-rich deposits (Figure 2.31). (We look in some detail at the dissolving of solids in liquids in Section 2.3.)

Figure 2.29
Settling of minerals in magma.

The result is that instead of an even distribution of minerals through the Earth's crust, there are areas where minerals are concentrated. Such concentrations can greatly affect the prosperity and political significance of those countries in which they are found.

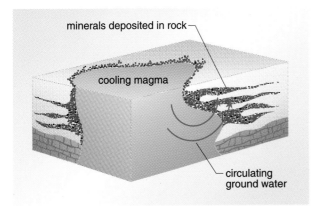

Figure 2.30
Deposition of minerals from circulating ground water.

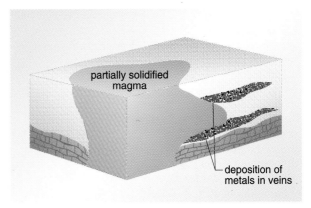

Figure 2.31
Sometimes, the high pressure of the magma opens cracks in the solid surrounding rock and the metal-containing minerals are deposited in a network of veins.

The map in Figure 2.32 shows the distribution of major mineral deposits over the Earth. Compare this map with the one in Figure 2.33 which shows the areas of the world that are subject to volcanic activity.

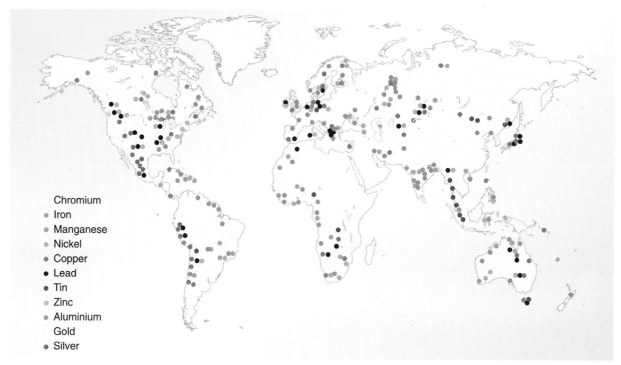

Chromium
Iron
Manganese
Nickel
Copper
Lead
Tin
Zinc
Aluminium
Gold
Silver

Figure 2.32
World-wide distribution of mineral deposits.

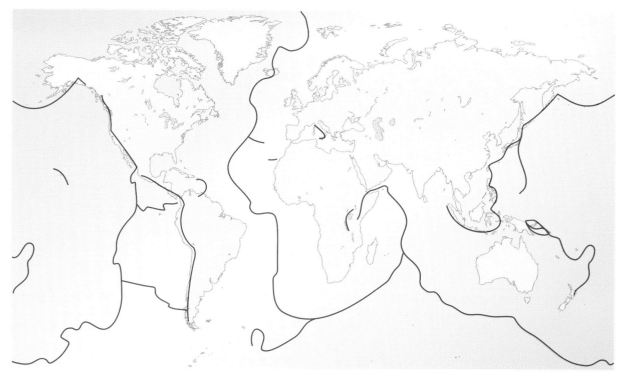

Figure 2.33
Areas of volcanic activity over the globe.

It does seem that there is some connection between the two maps. In areas where molten rock is close to the Earth's surface, many minerals are to be found. Minerals are continually separating from magma, and it is no coincidence that the volcanic regions of the Cascade Range of mountains which run north–south on the western side of North America are associated with extensive mineral deposits (Figures 2.34 and 2.35).

Figure 2.34
Crater Lake, in the Cascade mountains, Oregon, USA.

Figure 2.35
The Bingham Canyon copper mine to the south-west of Salt Lake City, Utah in the USA is the largest excavation on Earth. It reaches a depth of 800 m over an area of 7.5 km^2 and has involved the removal of 7×10^9 tonnes of material. If all this material were put in trucks, they would stretch nose to tail from the Pacific to the Atlantic across the USA. This excavation is reportedly one of the few indications of habitation of the Earth observable from the Moon.

Weathering and erosion of rocks is also important in concentrating minerals. Minerals are freed as rocks are broken down by wind, water and ice and are 'winnowed' according to their particle size, hardness and density. Denser minerals may be deposited from rivers on the outside of bends where the current is slacker. The deposits of gold in the days of the Klondike were formed in this way as were the tin deposits of Malaysian rivers and the titanium-rich sands of some Australian beaches. In Chapter 4, we shall look in detail at how metals are extracted from minerals.

At this point, it is worth undertaking a short experiment which illustrates how water can bring about the separation of different materials.

Experiment 2.1 Separation of materials with water

2.3 Dissolving and concentration

The basic ideas of solubility were introduced in Book 1 but we now need to go further. However, we shall start by briefly recapping that discussion. Sweetening tea and coffee with sugar is an everyday activity for many. Let us think about what might be happening as sugar dissolves in tea. A single teaspoon of sugar added to a cup of tea dissolves. A second teaspoon of sugar would also dissolve.

■   What would happen if more sugar were added, maybe as much as 50 teaspoons?

■   You will probably have had no reason to carry out such an experiment. Some of the sugar would dissolve and some would remain as an undissolved solid on the bottom of the cup. The result of a similar experiment is represented in Figure 2.36.

Figure 2.36
(a) Addition of 10 g sugar to 100 g water (left) to give a solution (right).
(b) Addition of 150 g sugar to 100 g water (left) to give a solution (right) and undissolved solid.

There is a limit to the amount of sugar that can be dissolved in water at a particular temperature: any additional sugar simply remains as a solid and collects on the bottom of the container and the solution is said to be **saturated**. This observation is quite general. A similar experiment using table salt (sodium chloride) instead of sugar would show that about 36 g of salt would dissolve in 100 g of water at 25 °C and the rest would remain as undissolved solid. Solubility can be measured as the quantity of one substance that will dissolve in a particular quantity of another. The solubility of sucrose (the chemical name for household sugar) in water at 25 °C is about 115 g per 100 g of water.

(a)

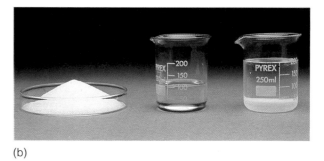
(b)

Solubility depends on both the solute and on the solvent and also on the temperature. Table 2.4 shows the solubility of a number of solutes in water and other solvents.

Table 2.4 Solubility of substances in a range of solvents. (In each case, solubility is expressed for 100 g solvent at 25 °C; for simplicity, values are given to the nearest whole number.)

Substance	Solubility in water	Solubility in ethanol	Solubility in hexane
sucrose	115 g	insoluble	insoluble
paraffin wax	insoluble	insoluble	10 g
potassium iodide	147 g	1 g	insoluble
sodium chloride	36 g	insoluble	insoluble
sodium nitrate	92 g	6 g	insoluble

Solubility varies with temperature (Figure 2.37). For example, the solubility of sodium nitrate doubles when the temperature of water is increased from 25 °C to 100 °C.

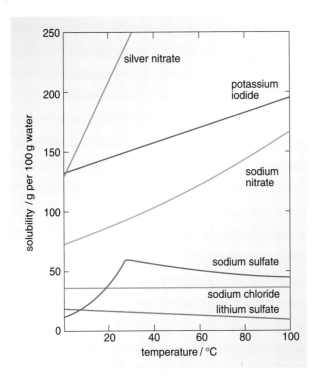

Figure 2.37
Variation of solubility of
some metal compounds in
water with temperature.

Notice that there is considerable variation in the way that solubility changes with temperature among the compounds featured in Figure 2.37. The solubility of silver nitrate increases very rapidly with increasing temperature as do the solubilities of both sodium nitrate and potassium iodide. The solubility of sodium chloride increases only slightly and that of lithium sulfate actually decreases as the temperature increases. The behaviour of sodium sulfate is rather curious in that it increases until a temperature of about 25 °C is reached and then gradually decreases. It is not possible to predict solubility but many compounds do show some increase in solubility with a rise in temperature.

In Table 2.4, the solubility of sodium chloride in water at 25 °C is given as 36 g sodium chloride per 100 g water. This implies that if 36 g sodium chloride is added to 100 g water at 25 °C, all the sodium chloride will dissolve to give a solution of sodium chloride in water. If less than 36 g sodium chloride is added, then again a solution is formed.

▨ What would you expect to happen if say 40 g of sodium chloride were added to 100 g water?

▪ A solution would be formed in which 36 g sodium chloride dissolved in the water and the remainder would not dissolve (Figure 2.38).

The solubility of sodium chloride is 36 g per 100 g of water at 25 °C. Both solutions in Figure 2.38 are saturated. A natural illustration of a saturated solution is the Dead Sea which is fed by the River Jordan. The water is saturated in sodium chloride as well as containing other dissolved minerals. The solution is more dense than water and this makes it possible for people to float effortlessly on the surface (Figure 2.39).

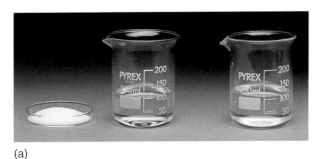

(a)

(b)

Figure 2.38
(a) Addition of 36 g sodium chloride to 100 g water (left) to give a solution (right). (b) Addition of 40 g sodium chloride to 100 g water (left) to give a solution and undissolved solid (right).

Figure 2.39
The Dead Sea is saturated in sodium chloride and other compounds. The density of the solution is greater than that of pure water and this enables bathers to float with ease.

Solubility is the mass of solute that will dissolve in a particular mass of solvent at a specific temperature to give a saturated solution. Once we have a solution, it is useful to be able to express the quantity of solute in a particular volume of that solution, that is the **concentration** of the solution. The concentration of a solution can be defined as the quantity of solute that is dissolved in a *certain volume of that solution*. Mass concentrations are usually specified in terms of the mass of solute per litre of solution.

▨   From the data in Table 2.4, do you think it is possible to specify the concentration of a saturated solution of sodium chloride in water at 25 °C?

■   A first thought might be that there is no problem. The data in Table 2.4 indicate that 100 g water will dissolve 36 g sodium chloride at 25 °C. If we take the density of water to be $1\,g\,cm^{-3}$, then there is 36 g sodium chloride per $100\,cm^3$ of water. But is the concentration of sodium chloride 36 g per $100\,cm^3$ solution? To say this, we should need to make the assumption that when 36 g sodium chloride dissolves in 100 g water, the volume of the resulting solution is exactly $100\,cm^3$. A glance at Figure 2.40 suggests that this is not the case.

Figure 2.40
Making a saturated solution of sodium chloride in water at 25 °C showing the change in volume from the water (left) to the saturated solution (right).

This problem is quite general; it is not possible to predict what the volume of a solution is going to be when known quantities of solute and solvent are mixed. Experiment shows that the dissolution of 36.0 g sodium chloride in 100 g water gives a solution of $111\,cm^3$. Even this is a remarkable result. The total volume of solute and solvent (when separate) is less in this case than is the volume of the solution.

▨   What is the concentration of a saturated solution of sodium chloride in water at 25 °C?

■   A saturated solution contains 36.0 g sodium chloride in a total volume of $111\,cm^3$.

In $1\,cm^3$ solution, there is $\dfrac{36.0}{111}$ g sodium chloride.

In $1\,000\,cm^3$ solution, there is $1\,000 \times \dfrac{36.0}{111}$ g sodium chloride.

It follows that the concentration of the solution is $1\,000 \times \dfrac{36.0}{111}\,g\,l^{-1}$
This is close to $324\,g\,l^{-1}$.

Question 5 Calculate the concentration of the following solutions in units of $g\,l^{-1}$.
(a) 2.3 g sodium chloride in a volume of 50 cm^3 of solution.
(b) 2.1 kg sucrose in a volume of 700 cm^3 of solution.
(c) 0.03 g sucrose in a volume of 0.5 cm^3 of solution.

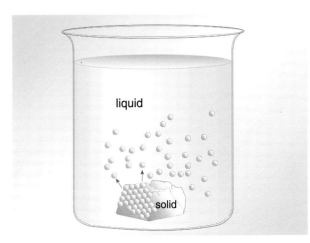

Figure 2.41
Solid dissolving in a liquid.

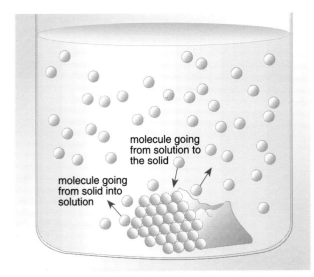

Figure 2.42
A molecular view of a saturated solution.

Let us look at solutions at the molecular level and see what happens when a solid dissolves in a liquid. When crystals of sucrose are added to water, the process of dissolving starts. Molecules close to the surface of the crystals break away and become dissolved molecules. This process goes on and, as new solid surfaces are exposed, more sucrose molecules pass into solution. This is the situation depicted in Figure 2.41.

There is, however, more going on than this. Sometimes a molecule in solution will approach the surface of an undissolved sucrose crystal and stick to it. The molecule passes from solution to the solid state. While the solid is dissolving, more molecules pass from the solid to solution in a given time than molecules move from the solution to the solid. The net effect is for the solid to dissolve.

As the concentration of sucrose molecules in solution increases, the rate at which molecules move from the solution to the solid also increases. Eventually, the two rates – molecules passing from solid to solution and from solution to the solid – become equal. At this point, there is no net dissolving occurring and we have a saturated solution in contact with the solid (Figure 2.42).

Overall, it appears that the situation is static. The volume of the solution does not change, the amount of undissolved solid does not change and the concentration of the solution does not change. There is an equilibrium. At the molecular level, however, the situation is far from static with sucrose molecules passing back and forth between the solid and solution states. We have *dynamic* equilibrium, a situation in which the rate of the material dissolving is the same as the rate at which the material moves from the solution to the solid.

Dynamic equilibrium is important in all areas of chemistry. You can get a feel for the idea by thinking about fish! Look at Figure 2.43a. The two fish bowls are connected by a tube which is wide enough for the fish to swim through. At the start of the experiment, goldfish are put into the left bowl. Some fish find their way along the connecting tube and reach the right bowl. To begin with, the movement of fish is one-way (Figure 2.43b). As the population of the right bowl increases, there is an increasing chance that a fish will swim in the opposite direction and move to the left bowl. Eventually, the populations of the bowls equalize with equal numbers of fish swimming left to right and right to left along the tube (Figure 2.43c). At this stage, on average, there is no net change in the populations of the two bowls but fish are continually passing back and forth. (This analogy assumes that these fish do not display any tendency to form shoals or family groups!)

2.4 Rocks

It is easy to overlook the contribution that rock hewn from the Earth has made, and continues to make, to our lives. The huge quantity of stone, brick and concrete that goes into the construction of a city the size of London is a reminder of the importance of building materials (Figure 2.44).

A walk in almost any part of the British countryside will reveal several different rock types representing just a small proportion of the wide variety of rocks that there are. Rocks vary greatly in colour and texture, and advantage is taken of this. For example, granite is tough and can be polished. Its attractive appearance finds use for the facings of prestigious buildings (Figure 2.45).

But just what is rock? You may recall from Section 2.2 that a mineral is a naturally occurring chemical compound of definite composition. Rocks are formed from mineral grains which have been cemented or crystallized together to form an interlocking mass. Some rocks may contain several different types of mineral grains but others, like limestone, comprise almost entirely a single type of mineral (Figure 2.46). The structures of rocks reflects their origins.

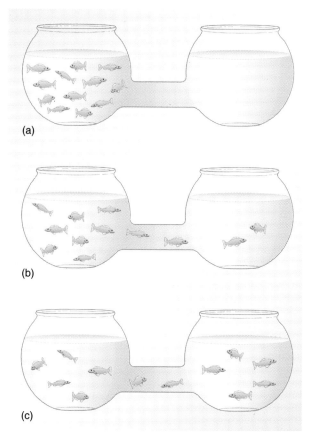

(a)

(b)

(c)

Figure 2.43
Dynamic equilibrium represented by the movement of fish between two bowls: (a) all the fish populate one bowl; (b) an intermediate stage; and (c) when dynamic equilibrium has been reached.

Figure 2.44
Stone buildings in Trafalgar Square, London.

Figure 2.45
A building faced with polished granite.

Figure 2.46
Limestone outcrops on a beach in southern Thailand.

In a rather immodestly titled book, *The Theory of the Earth with Proof and Illustrations*, published in 1785, the Scottish geologist James Hutton developed his rock cycle. His idea was that the Earth was not a static body but that its surface was constantly changing. The sequence of events is shown in Figure 2.47.

Figure 2.47
The rock cycle envisaged by Hutton. The cycle begins (1) with igneous rocks which originated in the molten mantle under the Earth's crust. Weathering of these rocks forms sediments (2) which become compacted and then are thrown up by Earth movements to form mountains (3). These rocks may be modified by heat from volcanic activity before being eroded once more as the cycle goes round again (4).

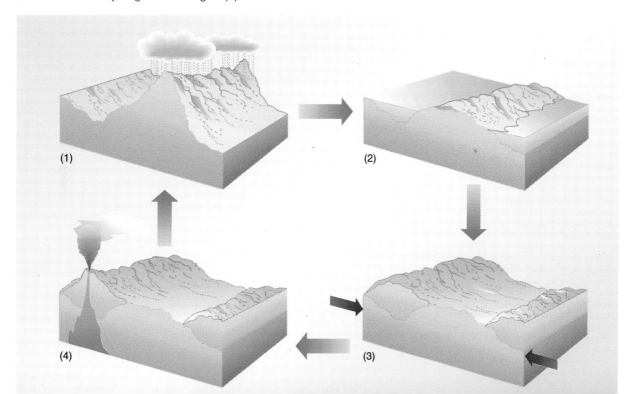

(a)

(b)

Figure 2.48
Bath Abbey (a) and the erosion of its stone carvings (b).

Sedimentary rocks such as limestone and sandstones have been very popular building materials. They are easier to work than the harder granites but are not so resistant to weathering and chemical erosion (Figure 2.48).

Native stone is a relatively expensive building material: quarrying, dressing and transport all adds to its cost. Cheaper and more adaptable constructional materials were needed and over the years brick, concrete, glass and steel have greatly influenced the design and structure of buildings. Even these materials have to come from the Earth and they require extensive processing before becoming useful. Weathering of some rocks gives clays which are the basis of the brick industry. Cement, which is the binding material in concrete, requires chalk or limestone in its manufacture. Glass is formed directly from silicon dioxide and steel is primarily the element iron obtained from mineral deposits of the metal oxide. We shall be looking at all these materials in Chapters 3 and 4.

2.5 Coal and oil

Coal is by far the most abundant fossil fuel on Earth and was burned to provide warmth by the Romans over 2 000 years ago. Oil and gas were first discovered by the Mesopotamian and Sumerian peoples of present-day Syria and Iraq. In the late 18th century, in the early stages of the Industrial Revolution, the main source of energy was from peat and wood. The use of the latter over the ages has resulted in massive deforestation of Britain. The

Figure 2.49
Rainforest in Queensland, Australia.

Figure 2.50
Formation of coal seams. Coal forms from the incomplete decay of vegetable matter which becomes dried and compressed. The effect of pressure and temperature slowly convert this to lignite and eventually to coal.

increasing need for energy to power the developing factories of the Industrial Revolution and the railway system saw coal become a major energy source. During the latter decades of the 20th century, there has been a move from coal to oil and gas. Oil and gas supplies will become increasingly scarce and expensive in the UK over the next 30 years but coal could supply our energy needs for some hundreds of years. The primary source of coal was the lush rainforests of prehistory, the like of which can still be seen today (Figure 2.49).

The forest floors became littered with partially decomposed fallen trees which were overlain by muddy silt from slow-moving rivers and then compressed. Further forest growth occurred and the cycle was repeated. The changing level of the land or changes in sea level, perhaps due to glacial melting, allowed this process to occur. Formation of coal is controlled by the conditions of temperature and pressure which in turn are related to the rate and depth of burial. At first, the softer tissues decomposed. As the temperature increased when the seam was buried to a greater depth, water was driven off and the proportion of carbon increased (Figure 2.50).

Coal formation goes through several stages: first with peat, then lignite and bituminous coal and, if the pressure and temperature are high enough, anthracite is the result. At each stage, the proportion of carbon increases. Anthracite contains over 90% carbon and often betrays its origins by showing the fossils of tree trunks and leaves (Figure 2.51).

Figure 2.51
Anthracite, with surface markings indicating its vegetative origin.

An interesting observation is that where fossils of tree trunks are found in coal seams, there is often an absence of seasonal growth rings. This would indicate that the early forests that were the source of the coal were located in the tropics where there is little distinction between the seasons. The distribution of the major coalfields of the world shown in Figure 2.52 indicates that there has been substantial drift of the continents, and many coalfields today lie far from the tropics.

Figure 2.52
The distribution of coalfields across the world.

Oil arises from a similar process except that its origin is not forests but minute plants of ancient oceans. When these plants died, they came to rest on the bottom of the oceans and were covered with silt. The nature of the decay and breakdown of these microplants with temperature is much more critical than it is with coal formation. Below 100 °C, a treacle-like mixture is formed in which there is much of the original plant material, and above 150 °C decomposition to carbon and the gas methane occurs. It is in this narrow window of only 50 degrees Celsius that oil is formed. In periods of high growth, the thick layer of decaying plants could give useful oil deposits. Slower rates of growth imply that any oil formed would be in lowish concentration and mixed in with the rock, as with the tar shales of Western Canada.

Under the right circumstances, oil can migrate into porous rock such as sandstone or limestone. If the rock is saturated with water, the oil, being less dense than water, rises until it meets an impermeable rock layer and

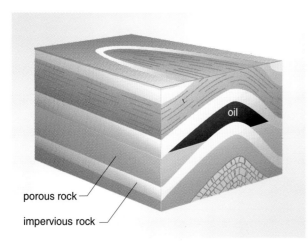

Figure 2.53
Oil trapped under a dome of impervious rock.

Figure 2.54
The petrochemical complex at Grangemouth, Scotland.

remains trapped (Figure 2.53). You can see that the chances of the right conditions for oil formation and its trapping in useful quantities require some fortune, thus explaining the highly localized distribution of oil.

In Britain, small seepages of oil have been known for centuries but had only been used for waterproofing boats and the like. In 1841, a Scottish chemist named James Young developed a process for the recovery of oily liquids from shales. It was this discovery that laid the foundations of what is now the huge petrochemical complex at Grangemouth in Scotland (Figure 2.54).

The accidental discovery (while drilling in search of water) of oil in 1859 in Pennsylvania, USA, by 'Colonel' Edwin Drake marked the beginning of the modern oil industry that has since changed the face of the world. It also marked the end, after just a few decades, of the Scottish oil shale industry.

Oil is not a pure chemical compound. It is a mixture of thousands of different substances most of which contain carbon and hydrogen. The numbers of potential isomers increase greatly with increasing numbers of carbon (and hydrogen) atoms in a molecule. The fact that crude oil contains molecules with up to several hundred carbon atoms gives an indication of the complexity of the mixture.

Much of the oil industry concentrates on separating this complex mixture and modifying the components for other uses. The major use of oil products is as a fuel: petrol and diesel for road transport, kerosene for jet engines, heavy fuel oils for ships and power stations all come from crude oil. Crude oil is not just valuable as a fuel but as a source of chemicals. Most polymers are derived directly from crude oil, and these materials are explored in Chapter 5. Fossil fuels as a source of energy are discussed further in Part 2 of this Book.

It is not just the structure of chemical compounds that is important but how one substance can be converted to another. There is a need to describe reactions rather more concisely than just by using words, and this is where chemical equations come in. As you work through this and the subsequent Books, you will see just what a useful tool the chemical equation is.

2.6 Chemical equations

A chemical equation can at first sight appear to be completely unintelligible, something apparently written in a foreign language. In many ways, that is just what the problem is. The chemical equation is written in a different language from that used for everyday communication. You have seen that symbols are used to represent the nature and numbers of atoms in formulas. This is done in chemical equations too but the symbolic language is carried further.

In the body, sugars react with oxygen to provide a source of energy. The overall process can be represented by the chemical equation

$$C_{12}H_{22}O_{11}(s) + 12O_2(g) = 12CO_2(g) + 11H_2O(l)$$

There are numbers, letters, subscripts, plus and equals signs, brackets all over the place. Let us begin our discussion with a simpler equation representing the reaction of methane with oxygen.

This reaction starts with methane and oxygen and these substances are converted to carbon dioxide and water. The first step is to write down what is present at the beginning of the reaction and what is there at the end. The reactants (starting materials) are written on the left and the products (final materials) on the right to give a verbal or word equation:

methane *and* oxygen   *go to*   carbon dioxide *and* water

reactants   products

Even the simple word equation carries much useful information in a relatively concise form but we can go further. You know that a formula tells us much more about a particular substance than does a name. The molecular formula of methane is CH_4 and this immediately indicates that the methane molecule comprises one carbon atom and four hydrogen atoms. The next step is to replace the names of the compounds in the equation with appropriate formulas:

CH_4 *and* O_2 *go to* CO_2 *and* H_2O

We can replace the word *and* by the plus sign, +. We shall eventually replace *go to* by the equals to sign, =, but we shall not do so at this stage. The reason for this will become apparent. So our reaction representation becomes

$CH_4 + O_2$ *go to* $CO_2 + H_2O$

Even at this stage of development, the relationship is becoming useful. Hydrogen atoms of the methane molecules in the reaction end up as part of the product water molecules and the carbon atoms end up as part of the product carbon dioxide molecules. It is important to note that the relationship is giving us information about the overall reaction. It does not say anything about how the reaction actually occurs such as the way in which bonds are broken and formed to give the final products.

■ Can you see a problem with the equation in terms of the *relative* numbers of reactant and product molecules indicated? You may find it helpful to use your molecular models to make a methane and an oxygen molecule. Then try to use just these molecules to make the product molecules, carbon dioxide and water.

■ It is not possible to do it. There is not enough oxygen and too many hydrogen atoms, as you can see from Figure 2.55.

Figure 2.55
The reaction of methane with oxygen represented by molecular models.

On the left, there is a molecule of methane and a molecule of oxygen. The carbon atom from methane appears on the right in the carbon dioxide molecule, which also contains the oxygen atoms. As there are just two oxygen atoms on the left (as an oxygen molecule), there is not enough oxygen to combine with the hydrogen atoms to form water molecules.

In going from the left to the right, there is a change in the numbers of atoms. In chemical reactions, atoms are neither created nor destroyed. There must always be the same number of atoms of each type in the products and in the reactants.

If you try the same exercise with the molecular models but this time start with *two* molecules of oxygen and one molecule of methane, you should find that there are sufficient oxygen atoms to form a carbon dioxide molecule and two water molecules. All the four atoms of hydrogen from the methane molecule can now be used to form two water molecules (Figure 2.56).

Figure 2.56
Models representing the equation for the reaction between methane and oxygen.

We need to be able to represent this information in our word equation. The relative number of each of the formula units that is required is written immediately in front of that formula in the equation:

$$CH_4 + 2O_2 \; go \; to \; CO_2 + 2H_2O$$

Where a number is not shown (such as in front of CH_4), you should assume (as with formulas) that there is a number one (1) implied.

Now that there is a balance in the numbers of atoms of each type, we can replace the words *go to* with an *equals to* sign to give

$$CH_4 + 2O_2 = CO_2 + 2H_2O$$

The *equals to* sign has a special meaning in a chemical equation. It indicates that the representation is an equation with the same numbers of atoms of each element on each side. That is, it is a **balanced chemical equation**. This is why we retained the words *go to* earlier, because at that stage, the reaction representation was not balanced. Chemical equations are just a shorthand way of representing information about a chemical reaction. They are favoured because they save lots of words and, with a little practice, you will be able to 'read' an equation just as easily as you can read prose.

Note that you cannot balance equations by changing chemical formulas. For example, methane molecules always contain one carbon atom and four hydrogen atoms, CH_4. It is only the relative numbers of formula units in the equation that are changed in the balancing process.

The skills of writing balanced chemical equations can be developed with practice. Let us go back to the reaction between sucrose and oxygen which was introduced at the beginning of this Section. Sucrose has the formula $C_{12}H_{22}O_{11}$. The reaction with oxygen can be written as

$$C_{12}H_{22}O_{11} + O_2 \; go \; to \; CO_2 + H_2O$$

To save having to write the words *go to* where the representation is not balanced, the symbol /=/ can be used. Where you see this symbol, you should be aware that there is no balance between the numbers of atoms in the reactants and the products. Anyhow, let us make a start in our balancing task:

$$C_{12}H_{22}O_{11} + O_2 \; /=/ \; CO_2 + H_2O$$

One approach is to consider each element in turn starting on the left. The sucrose molecule has 12 carbon atoms. There are no other carbon atoms on the left of the /=/ sign so we must arrange for 12 carbon atoms to be on the right. That means that there have to be 12 molecules of carbon dioxide on the right. Effecting this step gives

$$C_{12}H_{22}O_{11} + O_2 \; /=/ \; 12CO_2 + H_2O$$

Note that the 12 implies that there are 12 formula units of CO_2, that is a total of 12 carbon atoms and 24 oxygen atoms. The next element is hydrogen and there are 22 such atoms on the left. Of the products, only water contains hydrogen and each molecule has two hydrogen atoms. So 11 water molecules are needed on the right to give

$$C_{12}H_{22}O_{11} + O_2 \; /=/ \; 12CO_2 + 11H_2O$$

The last element is oxygen and there are just two atoms making up the single oxygen molecule on the left together with 11 oxygen atoms in the molecule of sucrose. On the right, there is a total of 35 oxygen atoms. Each of the 12 carbon dioxide molecules has two oxygen atoms giving 24 and there are another 11 oxygen atoms in the 11 water molecules. So, in addition to the 11 oxygen atoms in the sucrose molecule, a further 24 oxygen atoms have to be supplied by oxygen molecules on the left. A total of 12 oxygen molecules is required to give

$$C_{12}H_{22}O_{11} + 12O_2 \;/=/\; 12CO_2 + 11H_2O$$

Now all that has to be done is finally to check that the numbers of atoms in the reactants and products really do balance. On each side there are 12 carbon atoms, 22 hydrogen atoms and 35 oxygen atoms. The /=/ sign can now be replaced by the = sign to give the balanced equation

$$C_{12}H_{22}O_{11} + 12O_2 = 12CO_2 + 11H_2O$$

This balanced equation embodies a lot of information. It not only identifies the reactants and products and their respective formulas but indicates the relative numbers of each reactant molecule and the relative numbers of each product molecule. It also shows that carbon from the sucrose molecules ends up in carbon dioxide molecules, the hydrogen produces water and the oxygen in the product molecules comes both from the sucrose molecules and from molecular oxygen.

> **Question 6** Develop balanced equations for the reactions represented in (a) to (d).
> (a) $FeO + H_2 \;/=/\; Fe + H_2O$
> (b) $Fe_2O_3 + H_2 \;/=/\; Fe + H_2O$
> (c) $N_2 + H_2 \;/=/\; NH_3$
> (d) $CH_4 + Cl_2 \;/=/\; CH_2Cl_2 + HCl$

In our discussion of chemical equations, we have confined our attention to neutral molecules. Many reactions involve charged species, ions. The rules for balancing representations of reactions involving ions are exactly the same as they are for the reactions that we have explored so far. Look at one of the reactions that takes place when the gas sulfur trioxide SO_3 dissolves in water to give hydrogen H^+ ions and sulfate SO_4^{2-} ions:

$$SO_3 + H_2O = 2H^+ + SO_4^{2-}$$

The sulfate ion SO_4^{2-} can be regarded as a molecule in which four oxygen atoms are bonded to a sulfur atom but overall bearing a double negative charge which we represent with the superscript $^{2-}$. This is known as a **polyatomic ion**.

The equation is balanced: there are equal numbers of sulfur, oxygen and hydrogen atoms on each side. For ionic equations such as this, there is one more test that must be applied to ensure satisfactory balancing, that is the total charge on each side of the equation must be the same. In this instance, we have only uncharged species on the left. On the right, there are two hydrogen ions each with a single positive charge and a sulfate anion with a double negative charge. The total charge on the right is $2 - 2 = 0$. The total charge on each side is zero.

It is not necessary for the total charge on each side of an ionic equation to be zero; it is just that it must be the *same* on each side.

Question 7 Indicate which of (a) to (c) represent balanced chemical equations.

(a) $Ag^+ + Cl^-$ /=/ $AgCl$

(b) $BrO_3^- + Br^- + 6H^+$ /=/ $Br_2 + 3H_2O$

(c) $2Ag^+ + Sn$ /=/ $Ag + Sn^{2+}$

You may have noticed that in the original equation for the reaction of sucrose with water there were letters in brackets after each of the reactants and products. It is sometimes convenient to indicate the physical state of the substances in the equation. The symbol (s) after $C_{12}H_{22}O_{11}$ indicates that sucrose is in the solid state under the conditions of our reaction. The gaseous state of oxygen and carbon dioxide is shown by (g) and the (l) after water indicates the liquid state.

We shall use chemical equations extensively throughout this and subsequent Books. You will get plenty of practice with them so do not worry if it has taken you a little time to work through Questions 6 and 7.

AV sequence 'Talking chemistry' discusses the balancing of chemical equations, and this would be a good point for you to work through the sequence.

2.7 Water

There is no overall shortage of water on Earth. The problem is that it is not always of the right quality in the right place nor in the right state (Figure 2.57).

There are about 10^{18} tonnes of water on Earth (enough to cover Britain to a depth of over 4 000 km) but most of this is the salt water of the oceans. Only about 3% of all water is fresh and 80% of this is locked in the polar ice caps and not easily accessible. The largest quantities of liquid fresh water is found in the major freshwater lakes such as the Great Lakes of North America, Lake Balkhash in Kazakhstan and Lake Titicaca high in the Andes on the borders of Bolivia and Peru (Figure 2.58).

Not all rain and snow becomes surface water. Much is absorbed into the ground and taken in by the root systems of plants. Large amounts are rapidly evaporated back into the air.

The map in Figure 2.59 indicates the great variation in the amount of water falling annually in any one part of the world, and the effect of shortage can be seen dramatically in Figure 2.60.

Figure 2.57
The Matterhorn seen from the Valais region of Switzerland showing water in solid and liquid states. Water as a gas is invisible. The clouds near the mountain comprise tiny droplets of liquid water.

Figure 2.58
Floating rush islands on Lake Titicaca which are inhabited by Uros Indians.

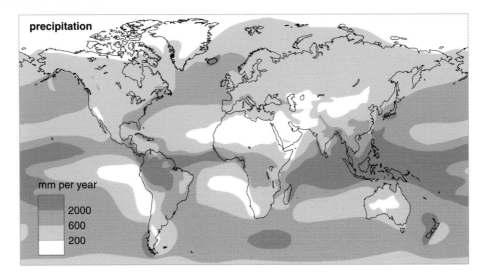

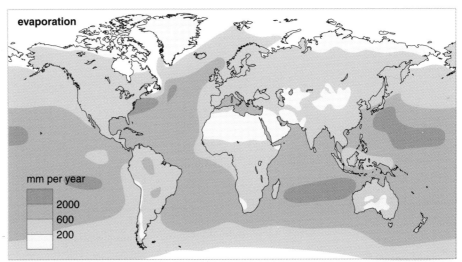

Figure 2.59
Worldwide mean annual precipitation and evaporation .

Water is being cycled continually, and much evaporation occurs over the oceans. Water-laden air moving from the sea and rising over coastal mountains cools quickly and rain (or snow) falls. Areas that have coastal mountains facing prevailing winds which have traversed the ocean tend to have high rainfalls. In Britain, the prevailing winds are from the west or south-west. The annual rainfall data in Table 2.5 (overleaf) are consistent with the topography of the country shown up on the map in Figure 2.61 (also overleaf).

Figure 2.60
An oasis in the Moroccan part of the Sahara desert.

Table 2.5 Mean annual precipitation data (as rain).

Location	Precipitation (cm equivalent for rain)
Aberdeen	76
Fort William	200
Borrowdale	329
York	62
Rhayader	248
Bedford	61
Norwich	67
London	60

Rain falling on land finds its way into streams and rivers and eventually back into the sea, and the water cycle shown in Figure 2.62 goes round again.

You may recollect that, for all its importance to life, water is a rather simple molecule comprising just three atoms: one oxygen and two hydrogen atoms. It is a most impressive solvent, and is capable of dissolving many different materials. Table salt (sodium chloride) dissolves in water, sucrose dissolves in tea (which is largely water), copper sulfate and iron sulfate dissolve in water. As we go through this Course, you will meet many different types of chemical compound and see just how remarkable are the solvent properties of water.

Figure 2.61
Topography of the United Kingdom.

Figure 2.62
The water cycle.

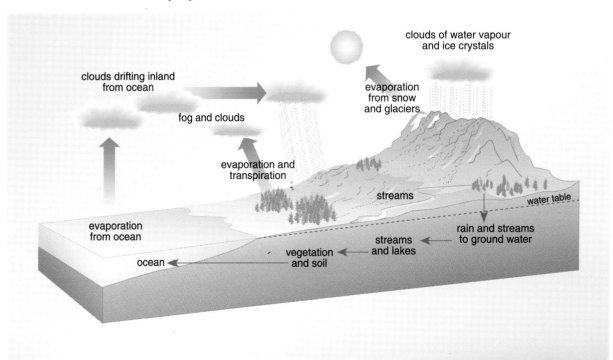

2.8 Ionic solutions and concentration

Many of the compounds that we have looked at so far comprise discrete molecules. In the solid state, sucrose molecules are packed together in an ordered array. When sucrose dissolves in water, all that happens is that the molecules move from the solid and become solvated by water molecules.

Ionic compounds are a little different, as you should recall from Book 1. Instead of being made up from individual molecules, the solid is an ordered array of ions. Sodium chloride is an example of an ionic solid: there is no such thing as a sodium chloride molecule. The solid is an ordered array of equal numbers of sodium ions Na^+ and chloride ions Cl^-.

■ Describe in simple terms what happens when solid sodium chloride is put into water.

■ When solid sodium chloride is put into water, it begins to dissolve. Sodium cations and chloride anions separate from the solid and go into solution as represented in Figure 2.63. It is important to realize that both the solid and the solution contain ions. Ions are not *formed* when the solid dissolves.

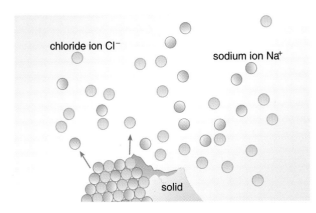

chloride ion Cl^-

sodium ion Na^+

solid

Figure 2.63
Dissolution of sodium chloride in water.

The concentrations of ionic solutions can be described using units of grams per litre $(g\,l^{-1})$ just as the concentration of a solution of sucrose was described earlier. Let's take a solution at room temperature which has a volume of one litre and contains 58.5 g sodium chloride. (The reason for taking this rather odd quantity will become apparent.) The concentration of sodium chloride in this solution could be described as $58.5\,g\,l^{-1}$. The solution contains a mixture of equal numbers of sodium and chloride ions.

■ What do you need to know to be able to express the concentration of the solution in terms of sodium cations and in terms of chloride anions?

■ You need to know the mass of the sodium cations and the mass of the chloride anions in the solution.

The key piece of information is the relative atomic masses of sodium and of chlorine: these values are 23.0 and 35.5 respectively. So the mass of one mole of sodium cations is 23.0 g and the mass of one mole of chloride anions is 35.5 g, a total of 58.5 g. Rather conveniently then, 58.5 g of sodium chloride contains 23.0 g of sodium cations and 35.5 g of chloride anions. The concentration of sodium cations is $23.0\,g\,l^{-1}$ and that of chloride anions is $35.5\,g\,l^{-1}$.

You may be concerned that we have assumed the mass of a chlorine atom is the same as that of the chloride anion (a chlorine atom which has gained an electron). A similar assumption is made when a sodium atom loses an electron to become a sodium cation. The electron has a mass of about

$\frac{1}{2\,000}$ the mass of the proton (about $\frac{1}{50\,000}$ of the mass of a sodium cation), so it would represent a very small correction indeed.

■   What is the concentration of chloride anions in 1.00 litre of solution containing 95.3 g magnesium chloride $MgCl_2$, a solid which dissolves in water to give magnesium Mg^{2+} and chloride Cl^- ions? (The relative atomic mass values for magnesium and for chlorine are 24.3 and 35.5 respectively.)

■   The empirical formula for magnesium chloride indicates that there are two chloride anions for every magnesium cation. The relative atomic mass of magnesium is 24.3, so of the mass of 95.3 g of magnesium chloride, 24.3 g is magnesium cations and the remainder, 71.0 g, is chloride anions. The solution then has a concentration of $24.3\,\mathrm{g\,l^{-1}}$ in magnesium cations and $71.0\,\mathrm{g\,l^{-1}}$ in chloride anions.

The equation representing the dissolution of sodium chloride in water is written as

$$NaCl(s) = Na^+(aq) + Cl^-(aq)$$

In this chemical reaction, ions that form solid sodium chloride become solvated by water molecules. Notice the symbol (aq). This is of a similar nature to the indicators of state (s, l, g) that you met earlier in our discussion of chemical equations and it shows that a substance is dissolved in water as an aqueous solution. For example, a sodium cation in solution is written as $Na^+(aq)$ and a chloride anion in solution as $Cl^-(aq)$. The symbol 'aq' implies that water is present although it is not formally represented as H_2O.

Expressing concentration in units of $\mathrm{g\,l^{-1}}$ is perfectly convenient for many purposes but for the chemist there are shortcomings. With sodium chloride, we saw that a litre of solution containing 58.5 g sodium chloride would actually have 23.0 g sodium ions Na^+ and 35.5 g chloride ions Cl^-. In chemical reactions, we are concerned with how many atoms, ions or molecules there are in our system and how many of one react with how many of another. Our solution of sodium chloride will contain equal numbers of sodium cations and of chloride anions but that certainly is not suggested by the mass of those ions in solution. What is needed is a way of expressing the relative *numbers* of ions in solution.

You may recall from Book 1 that the Avogadro constant indicates the number of entities that there are in one mole. This idea is perfectly general and applies to atoms, molecules or to ions which happen to be our focus at the moment. (Remember that the Avogadro constant is $6.02 \times 10^{23}\,\mathrm{mol^{-1}}$. That is, one mole of anything has 6.02×10^{23} entities.)

■   What is the mass of one mole of sodium cations and how many cations are there?

■   Sodium has a relative atomic mass of 23.0, so one mole of sodium cations has a mass of 23.0 g and contains 6.02×10^{23} cations.

The relative atomic mass of chlorine is 35.5 so a solution containing 35.5 g chloride anions has 6.02×10^{23} chloride anions. Now we can state that the litre of solution containing 58.5 g sodium chloride contains 6.02×10^{23} sodium cations and 6.02×10^{23} chloride anions. Or to put it more simply, the solution contains one mole of sodium cations and one mole of chloride anions. The mole allows us to compare directly the relative number of ions, atoms or molecules. Using the idea of a mole, it is easy to see that this solution contains equal numbers of the two ions. The concentration of this solution of sodium chloride in water is expressed as $1.00 \ mol \ l^{-1}$. This is known as the **molar concentration** as distinct from just concentration which is generally expressed in terms of $g \ l^{-1}$.

> **Question 8**   Calculate the molar concentrations of each of the ions in $20 \ g \ l^{-1}$ solutions of the following ionic compounds:
>
> (a) KBr, which dissolves to give K^+ and Br^- ions;
>
> (b) Na_2S, which dissolves to give Na^+ and S^{2-} ions;
>
> (c) $FeCl_3$, which dissolves to give Fe^{3+} and Cl^- ions.

As you progress through this Course, you will find that the concept of the mole will help us find a simple route through the difficulty that we would otherwise have of dealing with such huge numbers of tiny atoms, ions and molecules. *This idea is discussed further in AV sequence 'Talking chemistry'*. The remaining Chapters in Part 1 of this Book take a close look at just how the Earth's resources have been used to provide humanity with a wide range of materials.

Summary of Chapter 2

Data can be represented in a number a ways: the table, pie chart and histogram in addition to the graph (Book 1). Each representation has particular virtues and weaknesses. Units in tables can be expressed using the slash notation, e.g. 'volume/cm³'.

The distribution of elements in the Earth is not uniform. There are different distributions in the major parts of the Earth's structure and within the crust. Minerals are concentrated in the Earth's crust by density differences, by differences in crystallization (melting) temperatures, by solubility differences and by air and wind. The Earth is a dynamic entity with rocks being cycled through the crust. Oil and coal are formed by the decay of living materials under specific conditions of temperature and pressure.

The atomic and molecular structure of many minerals is reflected in physical properties. Crystal structures can be represented by the unit cell. Atoms at the corners, edges and faces of unit cells are shared with adjacent unit cells. The chemical formula for a molecule represents the numbers of atoms of each type in the molecule, as does the molecular formula. For giant and extended structures, the empirical formula is used which simply expresses the relative numbers of atoms of each type, as does the molecular formula.

The solubility of substances cannot be predicted easily. Solubility is usually quoted in terms of the mass of solute that will dissolve in a particular mass of solvent at a given temperature. A saturated solution is a solution that contains the maximum quantity of a solute at a particular temperature. Concentration is the quantity of solute in a given volume (usually one litre) of solution. It is expressed in units of $g\,l^{-1}$ and molar concentrations in units of $mol\,l^{-1}$. When ionic solids dissolve, the ions separate and are free to move through the solution.

Reactions can be represented by balanced chemical equations which must have equal numbers of atoms on each side of the 'equals to' sign. Additionally, balanced ionic equations must have an equal overall charge on each side of the equation.

Question 9   Outline the main ways in which minerals are concentrated in the Earth.

Question 10   From the pie chart in Figure 2.64, construct a table and a histogram showing the materials in ascending order of abundance.

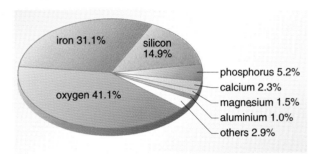

Figure 2.64
Pie chart showing the elemental analysis data for an iron ore sample.

Question 11   The unit cell for the mineral perovskite is shown in Figure 2.65. For the purposes of this question, assume that there are three types of ion in the unit cell. The calcium cation has a charge +2 and the oxygen anion a charge of −2. What is the charge on the titanium ion? (You may find it helpful to first obtain the empirical formula from the unit cell and then to calculate the appropriate charge for titanium.)

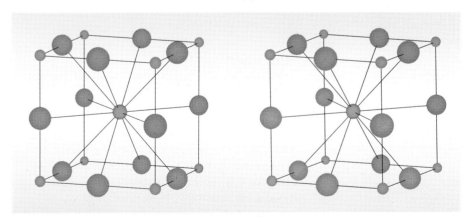

Figure 2.65
Unit cell of the mineral perovskite. The titanium ion is shown in blue, calcium ion in grey and oxide ion in red.

Question 12   When 27.8 g of the ionic compound magnesium iodide MgI_2 (which comprises the ions Mg^{2+} and I^-) dissolves in 100 g of water, the volume of the final solution is 105 cm^3. What is the molar concentration of the iodide ion I^- in this solution?

Chapter 3
Cement, glass and ceramics

Figure 3.1
Stone Age flint mine at Grimes Grave, Norfolk.

Figure 3.2
Stone Age flint tools.

The Stone Age (or maybe it should really be called the Rock Age), which arguably spanned the period from 10 000 BC to about 2200 BC, marked the transition from the nomadic hunter–gatherer society to a civilization where people were relatively settled and began to grow crops. It was not the quantity of stone that was used by this early society that gave the age its name. Dwellings of that time were often made from earth, wood and grass thatch, clothes came from animal skins or were woven from natural fibres like animal hair or kapok, and animal sinews and hide were used for bindings.

The distinguishing feature of this society was their use of stone for tools. Much stone has excellent properties of hardness and strength but cannot be made sharp. Even a hard stone like granite is not very useful for cutting tools. Flint is a glass-like silica material often found in areas of sedimentary rock (Figure 3.1). Its value is in the way that it breaks to give sharp edges (Figure 3.2). It was the improvement that was brought to cutting tools such as axes and arrow heads that made flint a highly prized commodity.

However, flint has its limitations. It is rather brittle and soon loses its keen sharpness. It cannot be sharpened except by chipping bits off to produce a new edge, so the tool eventually disappears. It is not easy to attach a handle to a piece of flint and this limited the size of tools that could be made. It was not until around 2200 BC with the advent of the Bronze Age that the technology was developed to produce metals. The important change that occurred was that materials were then being taken from the Earth and processed to produce new materials. The concern of this and the following Chapters in Part 1 is with materials that require significant processing after they have been won from the Earth.

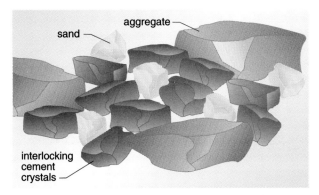

Figure 3.3
Structure of concrete.

3.1 Cement

Perhaps the most important building material today is concrete. Concrete is gravel or crushed stone with sand and water that has been mixed with cement and set into a solid mass. Cement forms an interlocking array of tiny crystals. It also forms chemical bonds with silicate materials of the crushed stone. It is this combination of physical and chemical features that gives concrete its strength (Figure 3.3).

The starting point in the manufacture of cement is limestone or chalk. These sedimentary rocks originated from the shells of sea creatures that lived in ancient oceans. When these shellfish, crustaceans and the like perished, their shells fell to the ocean floor and became buried and compacted. Subsequent earth movements and the drying out of inland seas resulted in the limestones of today. All limestone and chalk contains at least 90% calcium carbonate, $CaCO_3$. Crushed limestone, as aggregate, is used for constructions such as motorways and it is also important in the steel industry. There are three main types of limestone and their locations in England and Wales are shown on the map in Figure 3.4.

Carboniferous limestones are well consolidated, grey rocks and are found in northern England. Used in buildings and the dry stone walls of this area, they are also responsible for the spectacular scenery and cave systems of northern Derbyshire and Yorkshire (Figure 3.5).

Jurassic limestones are found further south in a band sweeping from Lincolnshire through the Cotswolds to Dorset. This stone is coloured by small quantities of iron oxides to give the distinctive pale-yellow look of Cotswold buildings (Figure 3.6).

Figure 3.4
The main areas of limestone in England and Wales.

Figure 3.5
Carboniferous limestone cliff at Malham Cove in the Yorkshire Dales.

Figure 3.6
Cotswolds house built from Jurassic limestone.

Figure 3.7
Cement works at Chinnor,
Buckinghamshire.

Chalk comes under the category of Cretaceous limestone and is to be found through the Chilterns and in Sussex and Kent where it forms the cliffs of Dover.

What we are concerned with here, however, is not limestone as a structural material but as a source of calcium carbonate. Chalk, where available, is the favoured raw material for the cement industry as it is generally over 97% calcium carbonate. The location of cement-making plants in England (Figure 3.7) follows the line of the limestone and chalk deposits down to Portland in Dorset after which Portland cement is named.

The first stage in the cement-making process is to heat calcium carbonate in a furnace so that the following reaction occurs:

$$CaCO_3(s) = CaO(s) + CO_2(g)$$

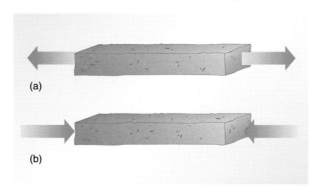

(a)

(b)

Figure 3.8
Concrete (a) in tension and
(b) under compressive
forces.

A temperature of over 800 °C is required. Calcium carbonate is not alone in the furnace: mixed with it is aluminium oxide Al_2O_3 and silica SiO_2 (often as an aluminosilicate rock). As the temperature of the furnace is raised further, the mixture partially melts and a cement clinker is formed. This contains a number of compounds but it can be regarded as being a mixture of calcium silicate and calcium aluminate. The cement clinker is ground into fine powder and is available in builders' yards all over the country.

One problem with concrete is that it is not a particularly good material in tension. If subject to a pulling force, it may fail catastrophically. However, it is extremely good under compression: it is not crushed easily and much of its use capitalizes on this property (Figure 3.8).

The tensile strength of concrete can be improved by reinforcement with steel bars. On major construction sites, you will often see networks of steel rods or mesh around which concrete is poured. The result is an effective and cheap building medium but it does suffer from a serious drawback. Water and air can penetrate the relatively porous concrete, and the steel reinforcement rods rust. This not only weakens the steel but damages the concrete. Rust takes up a greater volume than the steel from which it is formed and the expansion cracks the surrounding concrete. You may have seen the tell-tale rusty streaks on concrete structures that indicate internal damage (Figure 3.9).

Figure 3.9
Rust damage of reinforced
concrete.

3.2 Glass

A group of Phoenician sailors had come ashore. They were preparing to cook on the beach over a fire and, not finding suitable stones on which to put their cooking pots, used lumps of a mineral known as trona from their cargo. As the trona was heated in the fire and mingled with the sand, a strange liquid flowed. This [wrote the Roman historian, Pliny the Elder] was the origin of glass.

Figure 3.10
Glass decorated vase depicting Thetis reclining between Hermes and Aphrodite, ascribed to late first century BC Rome.

This almost certainly is not an accurate interpretation of history by Pliny but it does embody more than a shred of truth. Some of the oldest pieces of glass have been found in Egypt and in Mesopotamia (present-day Syria and Iraq) and date back over 4 000 years. The Egyptians and Romans were adept artists in creating all manner of intricately shaped and coloured glass objects (Figure 3.10).

Let's get back to the beach and the story of the Phoenician sailors. Trona is a mineral found in a number of areas of the western Mediterranean. Heating sand with sodium carbonate Na_2CO_3 (from trona) produces glass. This basically is the way that glass is still made today.

There is no evidence that glass was used for windows in the dwellings of those times: the need would have been minimal in the relatively warm climate of north Africa and the eastern Mediterranean. In Britain and northern Europe, the earliest examples of glass have been found in church windows and are no older than about 1 200 years (Figure 3.11). Here, the use of glass is decorative but it also serves as a transparent material to allow light into the interior of the church.

Most of the solid materials that we have met so far have not been transparent, but glass has the remarkable property for a solid in allowing light to pass through it with very little distortion. (Some plastics too are transparent) It is interesting to note that many liquids are transparent. The most common liquid, water, is certainly transparent and so is cooking oil, paraffin, alcohol, golden syrup and wine. Note that a liquid or a solid can be transparent and also be coloured (e.g. golden syrup or wine). A short experiment here will illustrate an unusual property of glass.

Figure 3.11
The church of Saint Michael in Stewkley, Buckinghamshire, dates back to the mid-12th century.

Experiment 2.2 Formation of 'glass'

Caramel is essentially sugar that has not crystallized and is transparent. It has another interesting property: if you leave it in a warm place, it slowly spreads out – it flows. Certainly, it seems to be a solid but it also flows very slowly. You might well wonder what this quality of appearing to be a solid yet being able to flow ever so slowly has to do with glass.

Cast your mind back to Book 1, which featured a stained glass window from the Cathedral at Chartres. During restoration work, it was noticed that the individual sections of glass were very thin at the top and thick at the bottom. Over five or six hundred years, the glass had flowed very slowly under gravity. Glass appears to be a solid but, like caramel, it does flow but even more slowly.

It may seem remarkable that the major ingredient of glass is sand. It is possible to heat silica until it melts and, if it is cooled quickly, a transparent material is produced. The problem is that silica does not melt until it reaches a temperature of over 1 600 °C and that is much higher than could be achieved with the technology available in ancient Egypt. By way of comparison, a coal fire in the home may reach 1 000 °C in the centre. However, a mixture of trona and silica melts at around 1 000 °C and forms a material which is transparent when it solidifies at about 650 °C.

The result of this reaction is a glass containing sodium (from the trona), silicon and oxygen but this glass slowly dissolves in water. This may not have been a major problem in the dry climate of Egypt but such a glass would not be useful for water containers, for instance. It did not take long before it was discovered that the addition of chalk or limestone to the trona and silica mixture gave a transparent glass that did not dissolve in water. Although silica and limestone are found widely over the world, the deposits of trona in the eastern Mediterranean enabled glass-making to begin there.

Heating silica with sodium carbonate results in the regular lattice of the silica being broken up. A chemical reaction takes place and the gas carbon dioxide is evolved. A very similar reaction occurs when chalk or limestone (calcium carbonate) is included in the melt. The resulting soda-glass is used today for window-glazing and for bottles and glasses.

The discovery of glass represented a significant step forward in the use of materials. Glass can be blown and moulded into a variety of shapes. It is not porous and can be used for containers for liquids. It is also very corrosion-resistant. Some of the most corrosive liquids like concentrated sulfuric and nitric acid are kept in glass containers. One problem with glass is that, although hard, it is fragile and shatters with a sharp blow. It is not just mechanical shock that will break glass either.

■ Take a look at the glass containers shown in Figure 3.12. What do you think may happen if boiling water is poured into the wine bottle, the beer glass and the kitchen measuring jug?

■ There is a fair chance that the wine bottle and beer glass will crack but the measuring jug should stay intact.

Figure 3.12
A wine bottle, beer glass and kitchen measuring jug.

The beer glass and wine bottle are made from soda-glass, which tends to expand significantly when heated. The hot water heats the inside of the bottle or beer glass. Glass does not transfer heat very quickly (it is a poor conductor of heat) and the inside of the bottle or beer glass expands while the cooler outside does not. The outer glass surface cannot withstand this pressure and so cracks. If you want to heat a soda-glass vessel, it is better to heat it gradually by filling the vessel with warm water, then replacing this with hotter water and finally adding boiling water. This gives time for heat to be transferred from the inside to the outside so the whole of the vessel expands at about the same rate and minimal internal stresses are created (Figure 3.13).

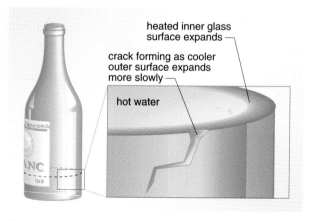

Figure 3.13
Stress caused by heating the inside of a soda-glass wine bottle.

▧   Which is more likely to crack when heated: a thick soda-glass tumbler or a thin one?

▪   The thick one.

It may seem strange that a thick soda-glass tumbler is more likely to break when heated than is one made of thin glass. With thin glass, heat is transferred more quickly from the inside to the outside than with thick glass. The result is a much smaller difference in the expansion of the inside and the outside and so the vessel does not break as easily.

So why is it that the measuring jug does not break although it is made of thick glass? The answer is that it is not made of soda-glass. The glass used is known as borosilicate glass which is similar to soda-glass but, as the name suggests, the element boron (in the form of boron oxide) is included during manufacture. Borosilicate glass, which is known under several trade names including Pyrex™, expands much less than soda-glass. It still does not conduct heat well but because the inside of the jug expands by only a small amount when boiling water is added, the stress is insufficient to crack the jug.

One way of overcoming limitations imposed by the fragility of glass is seen in laminated and toughened glass for car windows. By adjusting the rate at which the glass is heated and cooled in manufacture, it can be toughened and an internal layer of polymer holds the whole window in one piece should the glass be damaged.

Modifications to the properties of glass can be achieved by altering the structure at a molecular level. We shall look at just one aspect, the colour of glass. Ordinary window glass appears to be colourless but look at a piece edge on – it is distinctly green. The element iron is widespread in the Earth's crust and finds its way into many materials. Silica from which the glass was made contains small quantities of iron. The glass is made under conditions which result in the iron impurity giving a very pale green colour. Under different conditions, where there is more oxygen around, iron can impart a red–brown colour to materials. The colour of sand is due to oxidized iron and so is the colour of many clays and terracotta pottery. Copper is another metal that can give more than one colour to glass, either red or blue–green. Table 3.1 shows the effect that different elements included in glass can have on colour and other properties. A range of coloured glass objects is shown in Figure 3.14. Just how these colours arise will be discussed in Book 4.

Table 3.1 Effect of incorporated elements on the properties of glass.

Element	Effect
boron	low thermal expansion
chromium	green
cobalt	blue
copper	red or blue–green
iron	brown or green
manganese	purple
selenium	red
lead	increase in 'brilliance' to produce lead glass and crystal
	large quantities of lead in glass give it the ability to absorb X-rays and gamma rays

Figure 3.14
A range of coloured glass objects.

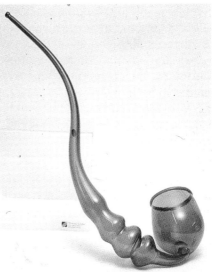

Experiment 2.3 shows how different colours can be introduced into glass.

Experiment 2.3 Coloured glasses

There is a 'dynamic' form of glass that has become quite commonplace over the past decade. Photochromic glass, glass that changes colour with light, has found widespread use in spectacles and in car windows, especially for cars in sunnier climates than that of the UK. Such glass is made by adding silver and copper compounds to molten borosilicate glass. As the glass cools, the silver and copper compounds form small crystals which are too small to be visible to the naked eye so the glass remains transparent. When this glass is exposed to sunlight (and especially to the ultraviolet radiation), an electron is removed from a copper cation and transferred to a silver cation to create a silver atom. Small clusters of silver atoms absorb visible light making the glass darken. When the source of ultraviolet radiation is removed, the silver atoms revert to silver cations and the glass becomes transparent again (Figure 3.15).

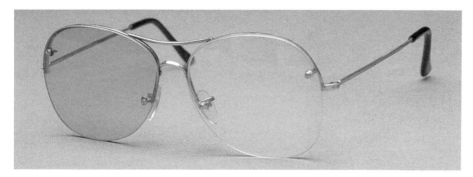

Figure 3.15
A pair of photochromic spectacles. Only the right-eye lens has been exposed to sunlight.

Glass is cooled rapidly during its manufacture so that there is not enough time for the structure to become sufficiently ordered to form crystals. Such crystals would prevent the glass from being transparent. Unlike a 'normal' liquid where the molecules can move relatively freely, there is very restricted mobility in glass. Even samples of 4 000-year-old glass still remain transparent. Under certain conditions, glass can become cloudy as a result of the formation of tiny crystals. This process of devitrification is similar to the effect that you sometimes see in a can of golden syrup or liquid honey that has been standing around for some time. Crystals form and the 'glass' of syrup or honey reverts to the crystalline form of the original sugar.

Devitrification of glass can also be seen with glassware that has been through many cycles in a domestic dishwasher. What looks like a cloudy film forms on the glass surface and, with subsequent washes, the glass becomes less transparent. Dishwasher powders contain materials that chemically scour the glass surface and encourage the growth of tiny crystals which render the glass opaque. Unlike a surface coating that could possibly be removed, the damage is to the actual structure of the glass and it is damaged forever (Figure 3.16).

Figure 3.16
A glass tumbler that has been washed many times in a dishwasher.

You may wonder why such materials are put into dishwasher powders if they can damage glassware. However, it is also the case that they are particularly good at dissolving fats and grease. There is a fine line between 'cleaning' and 'damage'.

Despite its long history, glass still has a bright future. It is already playing a leading role in telecommunications in the form of optical fibres.

Traditionally, cable communications have been built around pulses of electrons along metal cables. Copper, with its relatively low cost and high conductivity, was the metal of choice for both communication and electrical power transmission. Increasing need for communications in the latter part of the 20th century has made scientists look to other methods of cable communication that would be more effective than using copper.

Glass can be drawn out into very fine fibres similar to those that are sometimes used in loft insulation. The fibres are flexible, relatively robust and are able to transmit light. In the mid-1960s, the best fibres were able to transmit light for a few tens of metres. Today, fibres have been developed that can carry pulses of light over large distances and are able to transmit over 30 000 times as much data as an equivalent copper cable. Five optical fibre cables link Europe with the USA and another three connect Japan with the west coast of the USA. An undersea optical fibre cable to connect the UK to Japan is planned.

In a completely different arena is the use of glass in the disposal of radioactive waste from nuclear power stations. After the reprocessing of used nuclear fuel, there remains a highly radioactive and dangerous 'mud'. Radioactive isotopes of the metals caesium, cobalt and strontium in this waste will remain a hazard for many years. Molten glass is able to act as a solvent for these materials and is very effective in dissolving high-level nuclear waste. Particles of glass are mixed with the waste and heated to a temperature of 1 200 °C. The resultant shiny black glass (Figure 3.17) is sealed into steel containers with intended future burial in a 'geologically sound' area. The metal isotopes are much less likely to be leached into underground water if they are sealed in the glass.

Figure 3.17
A cylinder of glass of the kind used to store isotopes of the elements present in waste from nuclear fuel reprocessing.

3.3 Ceramics

Early ceramics were made from clays which had been hardened by heat. Such materials, moulded into cooking and eating vessels, provided humans with an early 'creative' medium. Pottery and earthenware represent the oldest ceramics but, with modern materials science, ceramics can be designed for such demanding uses as components of car and jet engines where they have to survive extreme mechanical and temperature conditions.

Pottery is just one type of ceramic. The three major ingredients of pottery are sand, feldspar and clay. As we have seen, sand is mainly silica. Feldspars are based on aluminosilicates incorporating metal ions such as

potassium and sodium. Clays have a wide range
of composition and arise from the weathering of
igneous rocks. Clay owes its properties to its
structure which comprises tiny platelets. When
the clay is wet, the platelets are lubricated and
can slide over one another allowing the clay to
be easily shaped (Figure 3.18).

The next stage is to allow the object to dry. As
water is removed, the platelets lock together and
the material becomes rigid. Firing in a kiln
removes any remaining water. The firing
temperature is kept somewhat below the
temperature at which the clay would melt in
bulk. The platelets partially melt at the edges,
where they are in contact with other platelets
and fuse together to give a durable product
(Figure 3.19).

Terracotta vessels (Figure 3.20) are porous and
if used to contain liquids the contents tend to
leak. This can be used to advantage in hot
climates: water in terracotta jars is kept cool by
the evaporation of water that slowly diffuses to
the outer surface. As this surface water
evaporates, the temperature is lowered in much
the same way that mountain walkers are more
susceptible to hypothermia in wet conditions
than in the dry.

Figure 3.18
Potter shaping a bowl.

For many purposes, porosity is a problem but it can be overcome by the
application of glazes to give a non-porous coating to the vessel. The clay
object is coated with a slurry of ground materials (such as silica) which
form a glass in the kiln. The particles melt together and effectively seal the
surface of the object. However, as Figure 3.21 admirably demonstrates,
glazes have had an aesthetic as well as a prosaic function from early times.

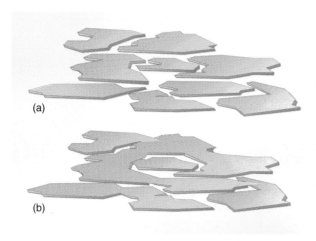

Figure 3.19
A representation of the platelets in clay (a) before and
(b) after firing, showing the fusing of the plates into a rigid
structure.

Figure 3.20
A terracotta bowl.

Figure 3.21
Glazed composition shabti
of King Seti I of the 19th
Egyptian Dynasty c. 1300
BC.

Box 3.1 Colour and glazes

Long before there were synthetic dyes and colouring agents, the craft workers of old produced a huge range of ceramic glazes. A visit to a museum will reveal the striking blues of Egyptian decoration, the red and black glazes of ancient Greece and, of course, the colour and artistry shown in the glazed porcelain of Imperial China.

The secret of a glaze is that the firing process must be at a temperature that is below the melting temperature of the clay of the object to be glazed. By adjusting the proportions of the ingredients of a glaze (silica, boron oxides, metal oxides), the melting temperature can be controlled. Firing is carried out so that the powdered materials of the glaze have the opportunity to melt and give a glass-like layer but that the time at elevated temperature is insufficient to damage the object being glazed.

As a glaze spends a minimum of time near its melting temperature, sometimes not all the particles have the chance to melt fully. The outcome is a surface coating that may contain unmelted raw materials or even air bubbles. These impurities can dramatically alter the appearance of the surface, making it opaque, translucent or

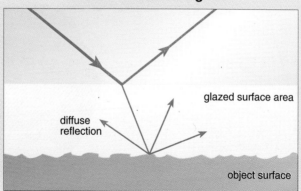

Figure 3.22
Interaction of light with a transparent glaze.

Figure 3.23
Lead-glazed bowl.

Figure 3.24
Matt glazing.

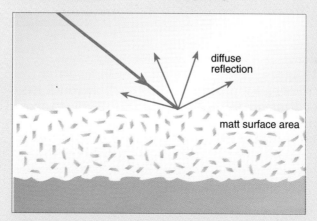

Figure 3.25
Interaction of light with a matt glaze.

transparent, smooth or matt. Control of composition of the raw glaze, particle size, firing temperature and the amount of air reaching the glaze during firing all form part of the glazer's art.

The appearance of a finished glaze is created by the interaction of light with the surface and the internal structure of the glaze. The so-called lead glazes give transparent coatings with a very smooth surface. Much of

the incident light is reflected without distortion from the surface of the glaze and there is a more diffuse reflection from the object surface on which the glaze lies. The mechanism of reflection is shown in Figure 3.22 and the result in Figure 3.23.

In contrast, a matt glaze has an uneven surface and light is reflected in all directions (Figures 3.24 and 3.25).

Between these extremes comes the celadon glaze. Such a glaze has a level of transparency but has been fired for a relatively short time. The surface is not completely smooth yet some light does penetrate into the interior. The body of the glaze contains unmelted silica crystallites and the reflections from these add to further reflections from the underlying object surface. The outcome is a glaze that is alive with subtle lights (Figures 3.26 and 3.27 overleaf).

Variety is increased by the range of colours that can be introduced. Compounds of many metals exhibit a huge variety of colours. Cations of chromium can give colours ranging from pink to green, cobalt gives deep blues, iron gives yellow, green, brown and even black and manganese is noted for

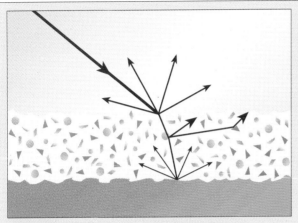

Figure 3.26
Interaction of light with a celadon glaze.

Figure 3.27
Celadon glazing.

purple. Firing can be adjusted so that metal compounds dissolve fully in the glaze to give an even colour or they can remain particulate to yield microcosms of colour. Figure 3.28 gives some indication of the range of colour that metals can impart.

Figure 3.28
Glazes coloured by metal compounds.

Ceramics are not just materials that have been used for pottery and decoration throughout the ages – they are making a niche in modern industry. A material that shows great promise is silicon carbide which has a structure similar to that of diamond. If it is manufactured carefully to avoid flaws in its structure, it proves to be a very resilient material being extremely hard, as tough as steel and as light as aluminium. Petrol engines are being developed with critical components made from silicon carbide and the related silicon nitride. These engines will be lighter and more efficient, and ceramic moving parts require minimal lubrication.

Modern engineering ceramics are to be found in many applications and are beginning to take over some of the traditional roles of metals. Space vehicles that have to re-enter the Earth's atmosphere are covered with lightweight, heat-resistant ceramic tiles, bone replacement is often now done with ceramics, and cutting tools in industry are often ceramic and not metal. The range of properties that can be designed into ceramics is enormous.

The limiting factor in the wider use of ceramics is that when they fail they do so catastrophically. Just think of the difference in dropping a pottery plate and a metal cooking pan on a hard floor. The metal pan may at worst suffer a dent; the pottery plate, however, does not deform but usually shatters. The breaking of ceramic materials arises from flaws in the structure such as the incorporation of foreign materials or occlusion of minute air bubbles.

The compound zirconium oxide ZrO_2 has found a niche as a component of many ceramics. When it is added to other materials such as aluminium oxide Al_2O_3, it bestows a useful advantage to the resulting ceramic material. Zirconium oxide can exist in two key crystalline forms. The different geometrical arrangement of the zirconium and oxygen ions in one form makes it about 4% more bulky than the other. Conditions of manufacture are adjusted so that the ceramic material contains the less bulky crystalline form. Should the material become damaged and a microscopic crack start to form, there is a change in the zirconium oxide to the bulkier form. This blocks the crack and the ceramic does not shatter. Methods of improving manufacture and composition are already resulting in ceramics that can easily withstand being hit by a hammer.

So ceramics are making a comeback. They represent some of the earliest materials used by humans and yet modern engineering ceramics are set to make huge inroads into the traditional role of metals in society. And it is to metals that we journey next.

Summary of Chapter 3

Cement is made from limestone ($CaCO_3$), alumina (Al_2O_3) and silica (SiO_2). When cement sets, it forms a rigid structure of interlocking small crystals. Concrete is strong under compression but not under tension. The tensile strength can be improved by reinforcement with steel but it is essential to prevent the steel from rusting.

Glass behaves as a solid under normal conditions but is actually a fluid. The range of properties and colour of glass can be adjusted by the addition of other materials. When heated, borosilicate glass expands much less than does soda-glass and is used in situations that would crack soda-glass.

Ceramics comprise small particles of solid oxides (usually) which are fired at a temperature that causes only partial melting so that the particles fuse together. These materials are able to withstand high temperatures and have densities lower than most metals. Ceramics are weakened by imperfections in manufacture. The introduction of zirconium oxide to a ceramic can help limit the propagation of microcracks. Glazes serve as a non-porous coating on a ceramic and as decoration. The colour, transparency and texture of glazes can be adjusted by small additions of metal compounds and control of firing times and temperatures.

Question 13 Suggest reasons why concrete is used extensively in the construction of buildings and bridges.

Question 14 The traditional vacuum or Dewar flask comprises a double-walled glass vessel. The glass used is extremely thin and consequently fragile. Why do manufacturers not use thicker glass to make the flask more durable?

Question 15 Indicate how you might achieve a glaze for a fired-clay vessel that has the following properties. The glaze must be non-porous; it must allow some light to be reflected from the surface of the vessel and some from the surface of the glaze; and the glaze should not be coloured in itself but it must incorporate small deep-blue colour centres.

Chapter 4
Metals

In this Chapter, we take a close look at the materials that have probably been the most influential in shaping society over the past two to three millennia – the metals. Go back to Figures 1.6–1.9 at the beginning of this Book and note the way that metals are being used in these pictures.

There is archaeological evidence in Britain of simple gold jewellery dating back beyond 2500 BC. Bronze brought about many changes to society but it was the development of iron technology that has had a continuing influence. The painting of 16th century Flanders (Figure 1.7) shows a range of roles for metals such as tools, horse shoes, and wagon wheel rims and nails for carpentry. There is a dramatic increase in the use to which metals have been put over the 3 000-year time-span represented by the pictures, certainly through to Victorian times (Figure 1.8). Here, metals seem to be very much part of the fabric of life and without metals and supplies of energy (which forms the focus of Part 2 of this Book), it is unlikely that the Industrial Revolution could ever have taken place. The mid-1990s picture (Figure 1.9) perhaps does not show a significant increase in the range of uses of metals compared with Victorian times. Other materials are now being used to replace metals.

You probably looked for uses of metals in these pictures without asking yourself just what is a metal. We all have an idea as to what is metallic and what is not, but let's pursue this distinction a little further.

4.1 What is a metal?

The fact that you have been able to identify some metals in the pictures suggests that you associate certain properties with metals. You would have little hesitation in distinguishing, say, the blade of a metal kitchen knife from the handle (which would probably be plastic or wood). What are these criteria and are there others that distinguish metals from other materials? A simple experiment may help here.

Experiment 2.4 Properties of metals

Scientists have tended to formalize metallic characteristics (as distinct from non-metals) by suggesting that metals are dense, lustrous (shiny), malleable (shaped by physical force) and good conductors of heat and electricity. Later in this Chapter, you will see that there are chemical criteria that help distinguish metals from non-metals.

Metals such as bronze, brass, pewter or stainless steel are not pure elements but alloys, blends of a metallic element with other materials. First, let us extend our criteria to a range of metallic elements. Data for some metals are shown in Table 4.1. The non-metallic element, sulfur, is also included in the table for comparison.

Figure 4.1
A selection of elements.
Clockwise from top right:
silicon, copper, sulfur,
aluminium, zinc, iodine and
mercury.

Table 4.1 Typical data for some common metallic elements and the non-metal, sulfur, at 25 °C.

Metallic element	Chemical symbol	Proportion in Earth's crust by mass/%	Cost/ £ kg^{-1}	Annual world production/ 10^3 tonnes	Melting temperature/ °C	Density/ g cm^{-3}	Heat conduction (1 is best, 11 is worst)	Electrical conduction (1 is best, 11 is worst)	Lustre of unprotected element (1 is high, 11 is low)
Aluminium	Al	8.3	0.82	7 800	660	2.70	4	4	8
Chromium	Cr	0.012	3.3	3 450	1857	7.19	7	9	3
Copper	Cu	0.006 8	0.95	4 750	1 083	8.96	2	2	5
Gold	Au	0.000 000 4	9 170	1.8	1 064	19.3	3	3	1
Iron	Fe	6.2	0.12	302 000	1 535	7.87	8	7	10
Lead	Pb	0.001 3	0.40	2 750	328	11.4	10	10	9
Magnesium	Mg	2.8	1.7	338.0	649	1.74	5	5	7
Silver	Ag	0.000 008	2 200	7.5	962	10.5	1	1	2
Sulfur	S	0.034	0.16	2 150	113	1.96	11	11	11
Tin	Sn	0.000 21	3.8	195	232	7.31	9	8	4
Zinc	Zn	0.007 6	0.70	3 945	420	7.13	6	6	6

The next task is to arrange the properties of the metals and sulfur in a 'league' table so that we can try to link the properties to the uses of the metals. Do this in Table 4.2 (overleaf) by inserting the symbols for these elements in the columns.

Table 4.2 Comparison of properties of some metals and the non-metal, sulfur.

Property	High ⟵								⟶ Low	
abundance in Earth's crust	Al	Fe	Mg	S	Cr	Zn	Cu	Sn	Pb	Ag	Au
cost	Au	Ag	Sn	Cr	Mg	Cu	Al	Zn	Pb	S	Fe
annual production	Fe	Al	Cu	Zn	Cr	Pb	S	Mg	Sn	Ag	Au
melting temperature	Cr	Fe	Ca	Au	Ag	Al	Mg	Zn	Pb	Sn	S
density	Au	Pb	Ag	Cu	Fe	Sn	Cr	Zn	Al	S	Mg
heat conduction	Ag									Pb	S
electrical conduction	Ag									Pb	S

Compare your completed Table 4.2 with the one at the end of Question 16 that we have already completed (Table 4.3).

> **Question 16** This question refers to the data in Table 4.3 (completed Table 4.2).
>
> (a) Which three metals are the best conductors of electricity? Why is just one of these three metals used much more than the other two?
>
> (b) Why is aluminium used extensively in the construction of civil aeroplanes?
>
> (c) Cans for food are made of iron with a thin coating of tin. How could a thin and even coating of tin be put on an iron can?
>
> (d) What connection is there between the abundance of a metal in the Earth's crust and its cost?

Table 4.3 Completed version of Table 4.2.

Property	High ⟵								⟶ Low	
abundance in Earth's crust	Al	Fe	Mg	S	Cr	Zn	Cu	Pb	Sn	Ag	Au
cost	Au	Ag	Sn	Cr	Mg	Cu	Al	Zn	Pb	S	Fe
annual production	Fe	Al	Cu	Zn	Cr	Pb	S	Mg	Sn	Ag	Au
melting temperature	Cr	Fe	Cu	Au	Ag	Al	Mg	Zn	Pb	Sn	S
density	Au	Pb	Ag	Cu	Fe	Sn	Cr	Zn	Al	S	Mg
heat conduction	Ag	Cu	Au	Al	Mg	Zn	Cr	Fe	Sn	Pb	S
electrical conduction	Ag	Cu	Au	Al	Mg	Zn	Fe	Sn	Cr	Pb	S

The non-metal, sulfur (S), is the poorest electrical and heat (thermal) conductor of the elements in Table 4.1 by some way. For example, it conducts electricity about 10^{22} times (that is $\times 10\,000\,000\,000\,000\,000\,000\,000$) less effectively than even the worst metallic conductor in Table 4.1, lead. Sulfur is such a poor conductor that it is not regarded as a conductor at all. It is an effective insulator being as good as the plastic insulation that surrounds electrical cables in the home. The metals are regarded as good conductors of both heat and electricity with only a factor of 13 separating the best electrical conductor in Table 4.1 (silver) from the worst (lead).

4.2 Gold, silver and copper

Most of the metallic objects recovered from archaeological sites are fashioned from either silver or gold and these were the first metals to be used by humans. Examples of gold ornamentation from Britain and ancient Egypt are shown in Figures 4.2 and 4.3. Gold has remained bright, shiny and lustrous whether it has spent millennia in the dry atmosphere of an Egyptian tomb or the cool damp of an early British burial site.

One obvious reason why gold was used is that it is a highly attractive metal that does not tarnish. It is ideal for jewellery and has the lure of exclusiveness. It is one of the rarer metals of the Earth's crust although this is not always the only criterion that affects the availability of materials. Gold was highly prized and could be used by the powerful as an indicator of social position. Gold also has a special place in religious ornamentation as evidenced by the Golden Buddha in Figure 4.4.

Figure 4.2
Iron Age torc made from an alloy of gold, silver and copper. Found near Snettisham, Norfolk, and believed to be 1st century BC.

Figure 4.3 (right)
Gilded Egyptian funerary mask probably belonging to a princess of the Middle Kingdom c. 1900 BC.

Figure 4.4 (far right)
The Golden Buddha at Wat Traimit, Bangkok. The solid gold statue, which is thought to date back to the 13th century, is 3 m tall and has a mass of 5500 kg. (At current market rates, the gold itself has a value of about US$75 million.)

Silver is another metallic element that stands the test of time in that it retains its metallic lustre and does not readily corrode. However, the main reasons that gold and silver were used in ornamentation by early civilizations, apart from their attractive appearance, is that these are the only two metals that are regularly found as pure elements in nature. Gold is sometimes found in veins of quartz and other minerals or in rivers where the flow of the water has washed away the rock. Only rarely is it found as the classic nugget (Figure 4.5) and much of the gold from the gold rush days of California and the Klondike was panned from river silt.

Small particles of the metal are separated from mud and rock by washing repeatedly with water (Figure 4.6). Eventually, the unwanted material is washed away and (with luck) a few glistening particles remain in the pan.

Figure 4.5
Gold embedded in the mineral calcite.

■ Look at Table 4.1. What property of gold makes this method of separation possible?

■ Of all the metals in Table 4.1, gold has the highest density at $19.3 \, \text{g cm}^{-3}$. By comparison, the density of rock is in the range $1.5\text{–}3.5 \, \text{g cm}^{-3}$ so when the mixture of rock and gold particles is agitated in water, gold tends to settle more rapidly.

Today, the process of extracting gold is much more difficult. The proportion of gold is so low in many goldfields that it cannot be extracted by simple physical separation and chemical methods have to be used.

We have made much so far of the resistance of gold to chemical attack but there in fact is a limited number of chemicals that will react with gold. Potassium cyanide is probably best known as an effective poison. In appearance, it is similar to table salt but just one-thousandth of a gram (a few grains) taken orally is sufficient to kill an adult human. Like sodium chloride, it is an ionic compound and is built from the potassium ion K^+ and the cyanide ion CN^-.

When gold is added to an aqueous solution of cyanide anions, an interesting reaction takes place. Solid gold dissolves in the solution as the $[Au(CN)_2]^-$ ion. This ion is known as a **complex ion** and comprises a gold atom with two cyanide (CN) units bonded to it. Overall, the charge is -1. Square brackets are sometimes used in the formulas of complex ions to indicate that the charge applies to the ion overall and is not necessarily associated with any particular atom of the ion.

Figure 4.6
Panning for gold in the American West of the 19th century.

Rock containing less than one-hundredth of 1% gold by mass is crushed and treated with cyanide ion. The soluble ion $[Au(CN)_2]^-$ is formed and the

Figure 4.7
Structure of the complex ion $[Au(CN)_2]^-$. The carbon and nitrogen atoms are black and blue respectively.

solution containing it is separated from the bulk of the solid rock. Addition of zinc metal to the solution liberates metallic gold. Cyanide ion is regenerated and can be reused:

$$2[Au(CN)_2]^-(aq) + Zn(s) = 2Au(s) + 4CN^-(aq) + Zn^{2+}(aq)$$

Unfortunately, the conditions under which this reaction is carried out in the field are sometimes not very carefully controlled. There is some escape of cyanide ion and consequent environmental poisoning.

Many metals form complex ions and molecules and these are often distinctively coloured. These colours form the basis of a convenient test for the presence of metals in rocks and were used by prospectors in the field.

Experiment 2.5 Testing for metals

We shall explore some other chemical methods of extraction of metals with copper and iron later.

■ From what you know about gold and silver, what can you infer about their chemical reactivity compared with iron?

■ Iron, unless it is protected by paint, oil or other covering, rusts away. Gold and silver do not. Iron is rather more reactive than is silver or gold.

Gold and silver do not react with air and water. The other metals in Table 4.1 do react with air and water and are not generally found in nature as the uncombined elements. Iron is usually found combined with oxygen and sometimes with sulfur. Winning iron from oxide ores is not easy and represents one of the major chemical industries.

Another property of gold that made it especially useful in jewellery is its relative softness. It is a very easy metal to beat into shape and can be hammered into sheets that are no more than 1 000 atoms thick. You saw in Book 1 that gold leaf is used for the decoration of buildings and early illuminated manuscripts. A little will go a long way: one cubic centimetre of gold hammered to this thickness would cover about 3–4 m², the area of a large double bed (Figure 4.8). In practice, gold is normally only beaten out to the extent that one cubic centimetre (1 cm³) covers about one square metre (1 m²). If it is hammered to be much thinner than this, it becomes very fragile and difficult to work (Figure 4.9, opposite).

Figure 4.9
Use of gold leaf decoration in a part of the Royal Palace in Bangkok. The photograph also shows the use of ceramics in decoration.

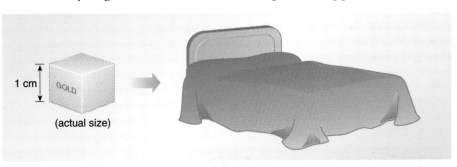

1 cm
GOLD
(actual size)

Figure 4.8
Area that could be covered by 1 cm³ of beaten gold.

Experiment 2.6 Simulation of atoms in a metal

Box 4.1 An atomic model for metals

Metal atoms pack closely together. Figure 4.10 shows bubbles on the surface of a liquid (each bubble can be thought of as representing a metal atom). Notice how most of the bubbles are in contact with six neighbouring bubbles.

But metals are three-dimensional. The layer of atoms represented by the bubbles would be covered by other layers and have layers underneath in a solid metal. This three-dimensional structure can be seen in the way fruit is sometimes stacked in market stalls (Figure 4.11).

Each metal atom in the three-dimensional structure is in contact with twelve others: six in its own layer, three in the layer above and three in the layer below. Virtually all metals adopt this kind of structure in which the atoms are neatly arranged in contact with neighbouring atoms.

Metals are good conductors of electricity: can this be explained by our model of metal structure?

Let's look at a single layer of metal atoms. Each metal atom comprises a nucleus with a positive charge and surrounded by electrons. A feature of metals is that they tend to lose electrons to form cations. This model can be regarded as a structure of regularly arranged metal cations in a 'sea' of electrons (Figure 4.12). The movement of one electron creates a hole that can be filled with a neighbouring electron, another electron moves into that hole and so on. The effect is the movement of electrons through the metal when an electrical potential is applied with a battery, electrical conduction (Figure 4.13).

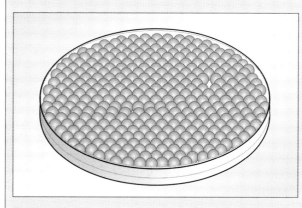

Figure 4.10
Bubbles packed together on a liquid surface. Note that there are small irregularities in the arrangement. Similar flaws in the structure of metals represent weakness areas and are sometimes responsible for the appearance of cracks.

Figure 4.11
Fruit stacked on a market stall.

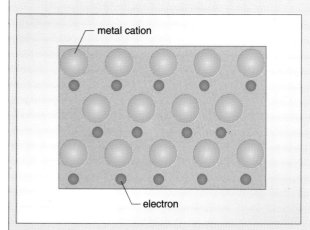

Figure 4.12
Metal ions in a sea of electrons.

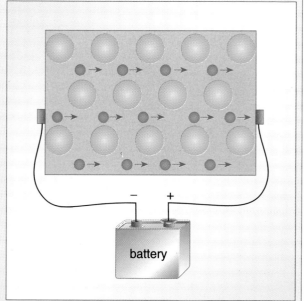

Figure 4.13
Conduction of electricity in a metal.

There is just one other metal that is occasionally found in nature in its elemental form and that is copper. In 1857, the largest single lump of so-called 'native' copper was unearthed in Minnesota, USA. It had a mass of 420 000 kg (about the mass of ten fully laden articulated trucks). Native copper is not often found and has no commercial significance today but it nevertheless must have had a significant impact on early civilizations, as did silver and gold.

Copper is arguably the most beautiful of all metals with an almost salmon-pink sheen. It is a soft metal and can be beaten into shape. Usually, the surface is dulled in air but a transparent lacquer can be applied to keep the surface shiny. Unfortunately, lacquering changes the colour of copper slightly. In most houses, the water pipes and central heating pipework is copper (Figure 4.14). Take some metal polish and give a small area a good rub – you may be surprised at the delicate hue.

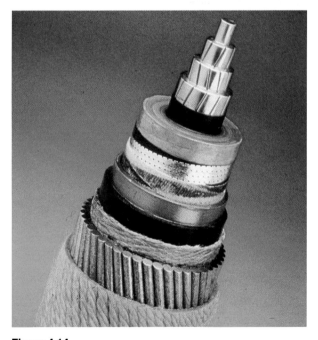

Figure 4.14
Some of the uses of metallic copper.

Early British silver and gold artefacts are usually found in good condition, but copper objects tend to be somewhat corroded. This observation is consistent with the fact that copper is usually extracted from ores by chemical processing. Most copper in the Earth's crust is chemically combined with other elements.

The usefulness of minerals as sources of metals depends on three main factors: the ease of extraction and purity of the mineral, the proportion of metal in the mineral and the ease with which the pure metal can be chemically produced. There are over 160 copper minerals known but only about a dozen or so are common. Many of these are brightly coloured, as you can see from Figure 4.15 and Table 4.4.

Figure 4.15
Native copper (lower centre), chalcopyrite (left) and malachite (right).

Table 4.4 Minerals of copper. (See below and the answer to Question 17 for missing entries.)

Mineral	Empirical formula	Proportion of copper by mass/%	Colour
native copper	Cu	100	reddish-pink
cuprite	Cu_2O	88.8	red
chalcocite	Cu_2S	79.9	dark grey
covellite	CuS	66.4	indigo
malachite	$CuCO_3$	57.5	bright green
chalcopyrite	$CuFeS_2$	34.6	golden yellow

In Chapter 2, we met the copper mineral chalcopyrite, $CuFeS_2$. To fill in the missing entry for chalcopyrite in Table 4.4, we need to calculate the proportion of chalcopyrite that is copper.

To carry out the calculation, we add together the relative atomic mass values in grams for the atoms in the empirical formula of chalcopyrite and this gives the formula mass.

For $CuFeS_2$, we have $63.5\,g + 55.8\,g + (2 \times 32.1\,g) = 183.5\,g$

A mass of 183.5 g chalcopyrite contains 63.5 g copper.

The proportion of copper by mass in chalcopyrite is $\dfrac{63.5\,g}{183.5\,g}$ = 0.346 or 34.6%.

Question 17 Calculate the proportion of copper by mass in covellite to complete Table 4.4.

Box 4.2 Copper mining in the English Lake District

The village of Coniston in Cumbria (Figure 4.16) is perhaps best known for its association with John Ruskin, the 19th century art critic, economist and social reformer. It was also the focus of an industry that had changed little for hundreds of years. Although mining of copper and lead ores ended here in the 1920s, there is much of the original workings and buildings still to be seen today. The whole of the valley running north-west from the village of Coniston is littered with mine shafts, spoil heaps and the stone buildings that were needed to house pumping equipment, mining tools and so on.

Figure 4.16
Location map of Cumbria.

Coniston lies close to the boundary of the volcanic Borrowdale rocks and the slates which transcend the Ordovician and

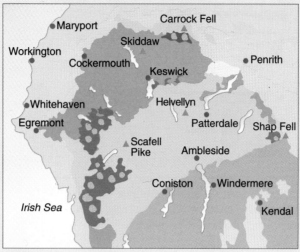

KEY

	rock type/group	geological Period
	younger rocks	
	limestone (sedimentary)	Carboniferous
	basal conglomerate (sedimentary)	
	sedimentary rock complex	Silurian
	Coniston limestone (sedimentary)	
	Borrowdale volcanic group	Ordovician
	Skiddaw Group	
	igneous intrusions	intruded at different time periods

Figure 4.17
Geology of the Coniston area.

Devonian Periods of about 430 million years ago (Figure 4.17). This time saw intrusions of magma into rocks and the separation of minerals, as was discussed in Chapter 2.

Mining began in this area at the beginning of the 17th century. Initially, it was possible to extract the copper minerals from very shallow excavations. Chalcopyrite, the major copper mineral of this area, was relatively easy to find with its distinctive golden appearance (Figure 4.18).

Figure 4.18
Chalcopyrite from the Coniston area.

Torrents of water rushing down from the mountains uncovered new veins. As excavations deepened, it became increasingly

difficult to extract the ore and to bale out the water that initially helped to uncover the veins. A solution was to drive a horizontal shaft into the hillside to connect with the lower part of a vertical shaft. These tunnels were created laboriously with a variety of iron tools and wedges at a time before blasting materials were available. Sometimes, fire was used to heat the rock and the contraction caused by sudden cooling with a dousing of water caused the rock to split. Some of the tunnels were extremely narrow, tapering from only 60 cm at waist height to a mere 20 cm at head level and often only 1.2 m high. Nevertheless, these horizontal 'drives' enabled accumulated water and ore to be removed with relative ease.

By the end of the 18th century, the Bonsor mine workings (Figure 4.19 overleaf) had reached a depth of 150 m and the ore was extracted through a tunnel of length of 200 m. A water wheel powered winch was used for winding and pumping. Changes in the scale of the extraction of copper ore can be seen by comparing the sketch of Bonsor mill with the Bingham canyon excavation in Figure 2.35.

Figure 4.19
Sketch of the Bonsor upper mill in Cumbria, 1850.

The fact that copper is occasionally found as the metal indicates that, like silver and gold, it is not particularly reactive. The converse is that when copper is in a chemical compound, it should be relatively easy to convert back to the metal.

We shall use chalcopyrite $CuFeS_2$ as an example but it is only one of several commercially valuable copper minerals. The first step is to heat the mineral in air. Most of the sulfur in the compound reacts with oxygen in the air and is removed as the gas sulfur dioxide, SO_2.

$$2CuFeS_2(s) + 3O_2(g) = 2CuS(s) + 2FeO(s) + 2SO_2(g)$$

At the same time, iron is converted to a solid metal oxide. The next step is to separate the mixture of oxide and sulfide, a process that takes advantage of the fact that copper sulfide melts at a much lower temperature than does iron oxide. Blowing air through the melt of copper sulfide is sufficient to remove the sulfur as sulfur dioxide and to leave metallic copper:

$$CuS(l) + O_2(g) = Cu(l) + SO_2(g)$$

The solid metal is known as blister copper because it contains bubbles of air and other impurities.

4.3 Tin and bronze

Although gold, silver and copper were available to civilizations several thousand years ago, these metals did not change the way that people lived in a significant way. Indeed, it was only the well-to-do that benefited from jewellery, decorations and vessels made from these metals. What was needed was a much harder metal, one that could be used for the heads of arrows and spears, for knives and other tools. Throughout the Stone Age, the best cutting tools had been made from flint.

The breakthrough came with the discovery of the metal, tin. Tin does not occur as the metal like gold and silver but often as an oxide as in the mineral, cassiterite. Just how this material, which looks like ordinary rock, was recognized as containing a useful metal is not known but the extraction of tin from cassiterite (Figure 4.20) represented the start of a metal extraction industry which has changed little in its chemistry to the present day.

The unit cell structure of cassiterite is shown in Figure 4.21.

Figure 4.20
Cassiterite.

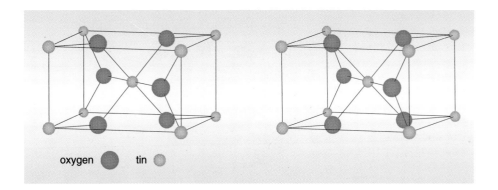

Figure 4.21
Stereostructure of the unit cell of cassiterite.

■ What is the empirical formula of cassiterite?

■ There are eight tin atoms at the corners of the unit cell each counting $\frac{1}{8}$ together with a tin atom at the centre, a total of two tin atoms. Four oxygen atoms on the cell faces each count $\frac{1}{2}$ and, together with the two oxygen atoms within the cell, there is a total of four oxygen atoms. The unit cell formula is Sn_2O_4 and the empirical formula is SnO_2.

Although it is not documented just how methods were developed for producing tin, it is likely that it started with a chance observation. Imagine a large cooking fire in a Stone Age village perhaps somewhere in Cornwall. The location is not an accident as this part of England had (and still has) deposits of cassiterite and other tin-bearing ores. Maybe there was a little cassiterite in the stone forming the base of the fire. The fire may have been used hundreds of times until someone raking through the ashes noticed a few, tiny silver-grey spheres. This would be metallic tin. Unfortunately, tin was not a metal that could be sharpened. The keenness of the edge was soon blunted and it turned out to be not as good as flint (although it could be shaped) for cutting tools.

Let's go back to the fire. The fuel would have been wood. As wood is heated, first water is boiled away and then the remaining carbon reacts with oxygen in the air to give carbon dioxide

$$C(s) + O_2(g) = CO_2(g)$$

In the centre of the fire and where the fire is in contact with the ground, the oxygen supply is limited. Here, there could be reaction between carbon and cassiterite in which liquid tin and carbon dioxide gas are formed. The reaction takes place in a number of stages but, overall, it can be represented by the equation

$$C(s) + SnO_2(s) = CO_2(g) + Sn(l)$$

However, the real value of tin in the progress of civilization is as a component of bronze. Again, we do not know just how it came about that bronze was made from copper and tin but it was probably a serendipitous observation. Often, tin and copper ores are found together and bronze may have been first produced in the normal production of copper.

Box 4.3 Don't rely on metal buttons!

Tin can exist in three different structural forms. Several elements have different structural forms which are known as **allotropes**. Diamond and graphite are allotropes of the element carbon. The physical properties of the three forms of tin are different and each form is stable over a particular temperature range. 'White tin' is the normal metallic form and this silvery metal is stable between about 13 °C and 161 °C. Between this range and 232 °C when tin melts, there is another form, rhombic tin (Figure 4.22).

'Grey tin', the form that is stable below 13 °C, crumbles easily to a grey powder.

Figure 4.22
Ingots of white tin.

Fortunately, it takes time for one form to turn into another. Tin that is used to plate metal food cans does not immediately become grey and powdery as soon as the temperature falls below 13 °C. Prolonged exposure to very low temperatures may bring about this change. Tin pest, as this change to crumbly grey tin is known, limits the use of metallic tin.

Napoleon's campaign to take Moscow in 1812 proved to be a disaster. The city was stoutly defended, and did not fall. As the French troops retreated, the Russian winter closed in. Extreme cold and a shortage of food led to the deaths of many men and animals. Buttons on the soldiers' uniforms were made of tin and the low temperatures caused a change to the grey form of tin. The soldiers' buttons just crumbled away and made the clothing even less able to withstand the rigours of the winter. While the change in the allotropic form of tin was not entirely responsible for the suffering and loss of life during this retreat, it is believed to have been a contributory factor.

The relative proportions of tin and copper can be varied over a wide range and the melting temperature of the alloy changes, as shown in Table 4.5.

Table 4.5 Composition and melting temperature of bronzes (copper–tin alloys).

Proportion of copper by mass/%	Proportion of tin by mass/%	Melting temperature/°C
100	0	1 084
90	10	1 005
80	20	890
70	30	755
60	40	724
50	50	680
40	60	630
30	70	580
20	80	530
10	90	440
0	100	232

■ Do the data in Table 4.5 indicate why bronze may have been a more useful material to early civilizations than copper?

■ As the proportion of tin in the alloy is increased, the melting temperature decreases. The melting temperature of copper is at the upper range of what can be achieved with a charcoal fire. By alloying copper with tin and lowering the melting temperature, bronze that can be melted in a fire is obtained. The liquid metal can be cast into sand or clay moulds making a whole new array of objects possible. Before, the only way to fashion copper was to beat it into shape.

Figure 4.23
Bronze Age artefacts.

If it were only for this ease of casting, it is doubtful that bronze would have made a significant impact on civilization. Certainly, there would not have been the long period in Britain, the so-called Bronze Age, where life was transformed through the properties of bronze (Figure 4.23). However, bronze turned out to be much harder than tin or the rather soft copper and it could be sharpened to a reasonably good cutting edge. Given, too, that the technology was available to cast the alloy into a range of shapes: hunting, warfare, farming, building, cooking and decoration were all transformed.

By experimenting, it was found that most useful properties were obtained with a tin content between 5% and 20%. The lowering of the melting temperature of copper compared with a 20% tin bronze (by 194 °C) is not great but it does bring the temperature down so that it is within the working temperature of a wood fire. However, this composition gives excellent hardness properties too. How is it that a relatively small amount of tin can increase the hardness of copper to such an extent?

The structure of metallic copper is shown in Figure 4.24 and comprises a regular array of copper atoms.

When copper is squeezed or beaten, layers of copper atoms can easily slide over one another. Look at Figure 4.25 which shows the inclusion of the larger tin atoms into the copper structure.

The larger tin atoms disrupt the layers of copper atoms and prevent them from sliding smoothly over one another. This, together with 'bonds' that form between copper and tin atoms, results in bronze being a much harder metal than is copper.

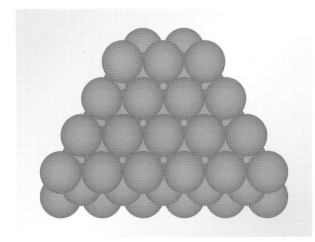

Figure 4.24
Structure of metallic copper.

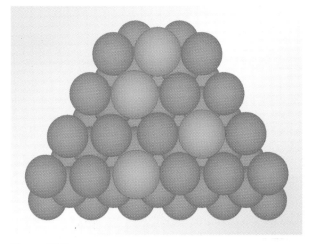

Figure 4.25
Structure of bronze.

Box 4.4 Gold in jewellery

Gold jewellery is rarely made from pure gold but from an alloy of gold and other metals, notably silver or copper. The effect is to harden the soft metal gold and make it more wear-resistant. Alloying also affects the colour of the metal. Pure gold is given a rating of 24 carats. If you have jewellery that is 9 carats you may not be pleased to learn that it is only 37.5% by mass of gold (9/24) and the rest is probably copper.

The low reactivity of

(a)   (b)   (c)

Figure 4.26
(a) Copper, (b) silver and (c) gold coins. The copper coin is Phoenician depicting Byblus, Emperor Macrinus. The silver and gold coins are Greek from Syracuse and Taras respectively.

gold, silver and copper is one reason why these metals have been used for so long in coinage (Figure 4.26). There would be little value in a coinage that slowly reacted to form compounds.

4.4 The Iron Age and iron

The Bronze Age represents a culture with a major dependency on a single working metal. True copper, silver and gold were known but they were not as useful as bronze (and, in the case of silver and gold, as available) for making cooking and storage vessels or axe and arrow heads. Comparison of the scene in the Bruegel painting (Figure 1.7) shows numerous changes, one of which is that many of the tools were made from iron. Iron is the element upon which the development of society was made from the beginnings of the Iron Age around 1000 BC and upon which the Industrial Revolution was founded. Even today, it is the most widely used metal. Why was it that iron ousted bronze as society's metal?

At first, it seems easier to find reasons why iron should *not* have superseded bronze. There are many examples of gold, silver, copper and bronze artefacts preserved from antiquity but there is hardly anything left that was made from iron.

Experiment 2.7 Rusting of iron

The fact that iron rusts is no great surprise as evidenced by the way that car bodies tend to corrode (Figure 4.27). In recent years, the protection of steel against rusting has been much improved and this is discussed in Section 4.5.3. Corrosion of many metals is much reduced in dry climates and it is not just chance that the US military mothballs their equipment in the Arizona desert (Figure 4.28).

Iron is widespread in the Earth's crust and, like tin, exists combined with oxygen. There are major deposits of iron oxide ores in Sweden, France, Australia, USA (particularly Minnesota) and in England. Iron ore deposits

Figure 4.27
Corrosion of the steel body of a car

Figure 4.28
United States Air Force planes mothballed near Tucson, Arizona.

of the east Midlands and Lincolnshire provided the base for the (now defunct) iron and steel industry of that area. It is not easy to persuade iron to release its hold over oxygen with which it is chemically combined. Like tin oxide, iron oxide has to be heated with carbon to win the metal, but the temperature required to carry out the reaction is higher and beyond that achievable with a wood fire.

Figure 4.29
Charcoal burning at the Weald and Downland Open Air Museum, Singleton, West Sussex.

However, two small developments that were to make a big difference occurred during the Bronze Age. It was found that if wood, instead of being burned directly on a fire, was partly burned in a limited supply of air, charcoal was formed. Charcoal is mainly carbon and proved to be a much better and hotter fuel than wood and made the manufacture of bronze so much more convenient. Indeed, the production of charcoal was a significant industry up to the early 19th century (Figure 4.29).

The temperature could be further raised by blowing air into the base of the charcoal fire with blowpipes. Even so, the real working limit of the fire of about 1 150 °C is well below the melting temperature of iron, 1 535 °C. Heating iron oxide in a charcoal fire through which air was blown resulted in metallic iron, but only because of a fortunate coincidence.

Carbon reacts with oxygen of the air and raises the temperature of the fire. In the centre of the fire, where there is little air, carbon begins to extract oxygen from iron oxide. As soon as the first iron is formed, carbon dissolves in it lowering its melting temperature. The effect on the melting temperature of iron of even small quantities of dissolved carbon is dramatic. The melting temperature of iron containing just 4% by mass of carbon is 1 150 °C, quite a fall from the 1 535 °C melting temperature of pure iron. So it was the development of Bronze Age technology that helped move culture forward to the Iron Age.

Iron ore does not come as pure iron oxide. Solid particles of rock are the major impurities in iron ore. The best ores may be up to 75% iron oxide but some of the useless dross tends to get incorporated into iron during the production process, the smelting. Soon, it was discovered that the addition of limestone to the ore gave a low melting temperature glassy slag which dissolved impurities. This slag was easy to separate from liquid iron as it floated on top. (Compare this with the discussion in Chapter 3 of the use of a glass as a 'solvent' for high-level radioactive waste from nuclear fuel reprocessing.)

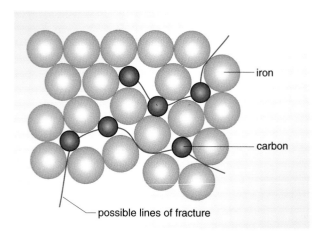

Figure 4.30
Structure of cast iron, showing weakness points.

Iron as produced from an Iron Age furnace contained about 4% by mass of carbon. Carbon atoms destroy the regular arrangement of iron atoms (Figure 4.30) making it hard but brittle. If cast iron is heated to about 850 °C, it becomes malleable and can be hammered into shape. Hammering iron at this temperature squeezes out any remaining slag and other impurities and also helps to expose the carbon so that it can be removed by reaction with oxygen in the air. This is the reason for the endless hammering in the fondly remembered poem *The Village Blacksmith* by Henry Wadsworth Longfellow:

> *You can hear him swing his heavy sledge,*
> *With measured beat and slow*

It was not simply to beat the metal into shape.

Removal of carbon makes iron much less brittle but also reduces its hardness, the very qualities one needs for tools, farm implements and weapons. The outcome was a compromise where some of the carbon was removed to give a resilient steel. It was found that if such steel (which is what iron with a low carbon content is) were cooled rapidly, the properties of toughness and strength are retained but the steel becomes hard and will retain a sharp cutting edge. The secret is to cool the steel rapidly, but simply plunging steel into cold water is not good enough. Water round the steel immediately boils and the resulting steam protects the steel and slows down cooling. The problem is that the water hardly touches the steel and heat is not conducted away quickly. It is not unlike the situation of a drop of water fizzing around on a layer of steam on an electrical hotplate. Years ago, urine was used as the tempering liquid. Crystals of the compound, urea, in the urine are formed on the steel surface and break up the steam barrier. The properties of the metal are also modified by carbon diffusing into the surface layers when steel was heated in the blacksmith's fire.

4.4.1 Modern production of iron and steel

From Table 4.1, you can see that iron is the metallic element with the highest annual production by far. It is used for industrial machinery, vehicles, railways, structural engineering, 'white goods' in the home, etc. Almost everywhere you look, you will find iron in use.

- Why is iron such a widely used metal?

- Iron is difficult to win from its ores (compared with say copper) and it is a metal that corrodes easily. However, it does have useful properties of strength and it can be shaped by hammering and pressing. Also, ores of iron are relatively common and iron ore often occurs in high concentration in particular sites.

There are two ores of iron that are commonly used in commercial steel production. Both ores are oxides: haematite Fe_2O_3 and magnetite Fe_3O_4. There is another oxide of iron that has the formula FeO.

Question 18 Calculate the proportion of iron, expressed as a percentage, by mass in the three oxides: FeO, Fe_2O_3 and Fe_3O_4.

Box 4.5 Using chemical equations to calculate quantities of reactants and products

The idea of the mole gives an alternative interpretation to chemical equations.

$$CH_4(g) + 2O_2(g) = CO_2(g) + 2H_2O(g)$$

This equation, as we have seen, represents an overview of a chemical change in which molecules of methane and molecules of oxygen are converted to molecules of carbon dioxide and molecules of water. It can also represent a process involving one mole of methane molecules and two moles of oxygen molecules. The difference between the two representations is enormous, a factor of 6.02×10^{23}, yet there is just one equation.

Question 19 How many particles are there in the following, and what is the total mass?

(a) 1.00 mol of chlorine atoms.

(b) 1.00 mol of hydrogen molecules.

(c) 0.50 mol of water molecules.

(d) 1.50 mol of oxygen molecules.

(e) 2.00 mol of nitrate anions.

Using the mole, it is possible to calculate quantities of reactants and products from a chemical equation.

- Methane is the major component of natural gas. The average house in the UK burns about 1 000 kg of methane each year. How much carbon dioxide enters the atmosphere from this source?

- The equation for the reaction is

$$CH_4(g) + 2O_2(g) = CO_2(g) + 2H_2O(g)$$

Using the molar interpretation of chemical equations, we can say that one mole of methane molecules is converted into one mole of carbon dioxide molecules. The relative molecular mass of methane is 16.0 and that of carbon dioxide is 44.0. So the mass of one mole of methane molecules is 16.0 g and the mass of one mole of carbon dioxide molecules is 44.0 g.

The equation tells us that one mole CH_4 is converted to one mole CO_2.

So 16.0 g CH_4 is converted to 44.0 g CO_2

1.00 g CH_4 is converted to $\frac{44.0}{16.0}$ g CO_2

1 000 g CH_4 is converted to $1\,000 \times \frac{44.0}{16.0}$ g CO_2

1.00 kg CH_4 is converted to $\frac{44.0}{16.0}$ kg CO_2

1 000 kg CH_4 is converted to $1\,000 \times \frac{44.0}{16.0}$ kg CO_2

So, on average, each house puts a staggering 2 750 kg or 2.75 tonnes of carbon dioxide into the atmosphere each year.

Question 20 Calculate the following:

(a) The mass of sulfur dioxide produced from 50.0 kg copper sulfide (CuS) according to the equation

$$CuS(s) + O_2(g) = Cu(s) + SO_2(g)$$

(b) The amount of carbon required to convert 1 tonne (1 000 kg) of the iron oxide magnetite Fe_3O_4 to iron according to the equation

$$Fe_3O_4(s) + 2C(s) = 3Fe(s) + 2CO_2(g)$$

The first step in the industrial production of iron is to grind iron ore into pieces about 10 cm in diameter. This requires large quantities of energy. The ore is mixed with calcium carbonate (limestone) and coke (the source of carbon) and dropped into the top of a blast furnace in a continuous process (Figure 4.31). The object is to remove oxygen from iron oxide to give metallic iron.

At the temperature of the furnace, carbon has a greater affinity for oxygen than it does at room temperature. This is one reason why the reaction is carried out at a high temperature. Also, most chemical reactions occur more quickly the higher the temperature, and that helps to produce iron more quickly. At room temperature, you could place a lump of carbon in contact with a lump of iron oxide and nothing would happen.

The chemistry in the furnace is not a simple reaction but a series of reactions. Remember that the charge to the furnace is a mixture of carbon, iron oxide (say haematite) and calcium carbonate. There will almost certainly be a fair proportion of silicon dioxide as impurity in the iron ore.

Hot air is blown in at the base of the furnace where coke burns to give carbon dioxide:

$$C(s) + O_2(g) = CO_2(g)$$

This reaction produces much heat and as a result the temperature in this lower part of the furnace is raised to about 1 900 °C. As carbon dioxide rises up the furnace, it meets more coke falling from the top and there is a reaction to give carbon monoxide. This reaction actually absorbs heat and the temperature of the part of the furnace where this takes place is at the lower temperature of about 1 400 °C:

$$CO_2(g) + C(s) = 2CO(g)$$

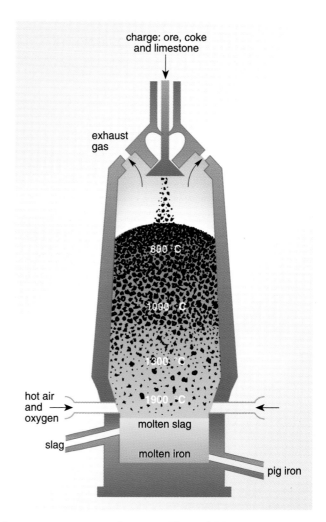

charge: ore, coke and limestone

exhaust gas

800 °C

1000 °C

1300 °C

1900 °C

hot air and oxygen

molten slag

slag

molten iron

pig iron

Figure 4.31
Blast furnace for the production of iron.

Here in the furnace there is no oxygen gas available but there is oxygen in the haematite ore. The affinity of carbon monoxide for oxygen is greater than that of iron at this point.

◼   Write an equation that represents the reaction of the iron oxide Fe_2O_3 with carbon monoxide to give iron and carbon dioxide.

◼   The unbalanced representation is

$Fe_2O_3(s) + CO(g) \; /=/ \; Fe(l) + CO_2(g)$

Balancing is achieved by noting that three carbon monoxide molecules are required to react with the three oxygen atoms in Fe_2O_3 and that two iron atoms will be produced.

The balanced equation is

$Fe_2O_3(s) + 3CO(g) = 2Fe(l) + 3CO_2(g)$

In the furnace, calcium carbonate decomposes to give calcium oxide and carbon dioxide:

$CaCO_3(s) = CaO(s) + CO_2(g)$

Remember that the major impurity of iron ore is rock containing SiO_2 and other oxides which have the rather high melting temperature of over 1 600 °C. Calcium oxide reacts with silica to give calcium silicate which melts at 1 250 °C:

$CaO(s) + SiO_2(s) = CaSiO_3(l)$

Silicon dioxide is removed as liquid calcium silicate from the lower part of the furnace. Note that the liquid calcium silicate floats on the liquid iron. The density of calcium silicate is about 2.5 g cm^{-3} and that of iron nearly 8 g cm^{-3}. Liquid iron is removed from a lower channel at the base of the furnace.

The product of the blast furnace is known as pig iron. It contains up to 5% carbon and is very hard but also brittle. Table 4.6 shows a typical composition for pig iron.

Table 4.6 Composition of pig iron.

Element	Proportion by mass/%
Fe	94.5
C	4.3
Si	0.7
Mn	0.3
P	0.1
S	0.05
others	0.05

It may seem surprising that a relatively low proportion of carbon can make iron so hard and brittle. This mass percentage is, however, a slightly misleading figure.

Question 21 Calculate the relative numbers of iron and carbon atoms in pig iron that contains 95% iron and 5% carbon by mass.

The answer to Question 21 indicates that there is about one carbon atom for every four iron atoms rather than the one carbon atom for twenty iron atoms which might be superficially suggested by a composition of 5% by mass carbon. This is certainly enough carbon to have a major effect on the structure of the iron.

The first step in the production of steel is to remove the impurities. Sulfur is the first impurity to be removed. The metal magnesium is added to molten pig iron and reacts with sulfur to form magnesium sulfide, MgS. The density of magnesium sulfide is $2.8\,g\,cm^{-3}$, much less than that of iron, so it floats to the surface and can be skimmed off:

$$Mg(s) + S(s) = MgS(s)$$

The molten, desulfurized pig iron is put into a brick-lined container, known as the converter, that will accommodate about 350 tonnes of pig iron. A water-cooled tube is lowered into it and oxygen gas blown through the molten iron with a spectacular display of sparks and flames (Figure 4.32). The impurities (carbon, manganese, silicon and phosphorus) are oxidized according to the following equations:

$$C(s) + O_2(g) = CO_2(g)$$

$$2Mn(s) + O_2(g) = 2MnO(s)$$

$$Si(s) + O_2(g) = SiO_2(s)$$

$$4P(s) + 5O_2(g) = P_4O_{10}(s)$$

Carbon monoxide escapes from the converter as a gas and is collected. It is burned to heat the air that is blown into the blast furnace. The temperature at which the impurities react with oxygen and the length of time that oxygen is blown into the converter are carefully calculated so that optimum conversion is achieved. Slag, containing oxides of silicon and other elements, and steel can now be poured separately from the converter as indicated in Figure 4.33. There is inevitably some conversion of iron to iron oxide in this process and this must be kept to a minimum. The contents of the converter are analysed at the beginning of the process and at several times during the oxidation so that steel is formed with just the right carbon content.

Wrought iron is relatively soft and can be hammered into shape. It has a carbon content of about 0.5%. Steel used for making car bodies and casings for refrigerators and washing machines would typically have a carbon content of 1–2% by mass.

Figure 4.32
Blowing oxygen through molten iron in the converter.

Figure 4.33
Pouring steel from a converter.

4.5 Reactivity of the metals

In Book 1, you learned that some metals are more reactive than others. In this Section, we shall explore this idea further and see what pattern there is in metal reactivity.

Go back to the beginning of this Chapter and list all the metallic elements that have been mentioned since then. Extract from the text any information that gives an indication of how reactive each metal is relative to other metals. Do this before you look at Table 4.7, where we have tried to pick out the main points.

Table 4.7 Comments on the reactivities of metals.

Element	Comment
gold	found as metal
silver	found as metal
copper	occasionally found as metal but more usually as sulfide or oxide
tin	found as oxide but can be reduced to metal with carbon at relatively low temperature
iron	found as oxide and reduced to metal with carbon (or carbon monoxide) at high temperature
calcium	the oxide is not reduced to the metal by carbon in a blast furnace
magnesium	reacts with sulfur preferentially to iron
manganese	oxidized by oxygen in the converter

From these observations, it is possible to arrange the metals in an order of reactivity. Now we do need to be cautious in that we are not really comparing like with like. For example, we have seen iron oxide reduced to metallic iron with carbon but sulfur was removed from impure iron by reaction with magnesium. Also, our observations are based on reactions at different temperatures. Nevertheless, metals do seem to follow an activity order irrespective of the other element with which they react.

Now, by filling in Table 4.8, try to create an activity order for the metals in Table 4.7, with the most reactive metal at the top.

Table 4.8 Reactivity of some metals in decreasing order (to be completed).

Metal

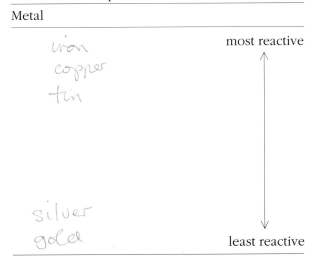

iron
copper
tin

most reactive

silver
gold

least reactive

Your table may show some variation from our completed version (Table 4.9). For example, from the information earlier in this Chapter it is not possible to decide where manganese should come relative to calcium and magnesium. However, you will probably agree that manganese is more reactive than iron because it is more easily oxidized (to manganese oxide, MnO) than is iron when impurities are being removed from pig iron.

This reactivity order holds well irrespective of whether we are looking at the reactions of the metals with oxygen, sulfur, chlorine or several other elements. For example, calcium reacts violently with chlorine but the reaction of chlorine with copper is much more gentle:

$$Ca(s) + Cl_2(g) = CaCl_2(s)$$

$$Cu(s) + Cl_2(g) = CuCl_2(s)$$

However, these observations represent only a rule of thumb and there are several factors that affect whether or not a reaction occurs and how quickly it occurs.

Another way of comparing the reactivity of metals is to look at reactions of metals in aqueous solution.

Table 4.9 Reactivity of some metals in decreasing order (completed Table 4.8).

Metal
calcium
magnesium
manganese
iron
tin
copper
silver
gold

Figure 4.34
Caesium stored in an evacuated ampoule.

Figure 4.35
The reaction of caesium with water.

Experiment 2.8 Reaction of copper cations with iron

If a piece of iron is put into an aqueous solution containing copper ions Cu^{2+}, a deposit of metallic copper is formed on the surface of the iron. What does this indicate about the reactivity of copper relative to iron?

Copper cations in solution have been converted to copper metal and to do this they have had to gain two electrons. The reaction can be represented by an equation in which an electron is denoted by 'e^-'. Just as the charge on an ion is shown by a superscript, the single negative charge on the electron is indicated in the same way:

$$Cu^{2+}(aq) + 2e^- = Cu(s)$$

The next question is from where have the electrons come? Iron dissolves in the solution forming positively charged ions, Fe^{2+}:

$$Fe(s) = Fe^{2+}(aq) + 2e^-$$

In forming ions, the iron atoms lose electrons and it is these electrons that are transferred to adjacent copper cations in the solution to form elemental copper. In this reaction, iron appears to be more reactive than copper, an observation consistent with Table 4.9. We know that iron reacts more readily with oxygen than does copper.

Similar experiments can be done with other metals and an activity series built up representing the tendency of metal ions to react with metals. The remarkable thing is the consistency of the activity series irrespective of whether they are developed from reactions in which one metal displaces another from solution or from reactions of metals with other elements.

There is a limit to the range of metals that can be used in these **displacement reactions**. Some metals, calcium is an example, react with water directly rather than displacing a metal from a solution:

$$Ca(s) + 2H_2O(l) = Ca^{2+}(aq) + 2OH^-(aq) + H_2(g)$$

With calcium the reaction is slow but metals such as potassium, rubidium and caesium react violently with water. These metals are so reactive that they have to be stored under oil or in a vacuum to protect them from oxygen and water (Figure 4.34):

$$2Cs(s) + 2H_2O(l) = 2Cs^+(aq) + 2OH^-(aq) + H_2(g)$$

During the reaction of caesium with water, enough heat is produced to ignite the hydrogen gas evolved and cause an explosion (Figure 4.35):

$$2H_2(g) + O_2(g) = 2H_2O(g)$$

An order of reactivity of metals can be developed that holds for a wide range of reactions. Such an activity series for a range of metals is shown in Table 4.10, with high reactivity at the top.

Table 4.10 Activity series for a range of metals.

Metal	Reactivity
caesium	high
potassium	
strontium	
calcium	
sodium	
magnesium	
aluminium	
manganese	
zinc	
iron	
tin	
lead	
copper	
silver	
mercury	
gold	low

4.5.1 Reactions of metals with acids

At the end of the 19th century, the Swedish chemist Svante Arrhenius (Figure 4.36) defined an **acid** as a substance that contains hydrogen and releases hydrogen ions $H^+(aq)$ in water. The covalent molecule, hydrogen chloride HCl dissolves easily in water. In this case, the process of dissolution results in a chemical reaction in which the covalent bond linking the hydrogen and the chloride atoms in the molecule is broken. Hydrogen cations and chloride anions are the products:

$$HCl(g) = H^+(aq) + Cl^-(aq)$$

Although hydrogen chloride does not contain hydrogen cations, it produces hydrogen cations in water and is therefore an acid according to Arrhenius' definition.

Many of the substances that can be classified as acids are of industrial importance (Table 4.11). It is interesting to note that none of these materials is generally available to the public. Their greatest use is in agriculture and the detergent and food industries.

A distinguishing feature of the chemistry of many metals is that they react with acids to form cations and liberate hydrogen gas. The addition of hydrochloric acid to zinc metal yields hydrogen and the ion, $Zn^{2+}(aq)$:

$$Zn(s) + 2H^+(aq) = Zn^{2+}(aq) + H_2(g)$$

You will notice that in the above equation, we do not include the other ion in hydrochloric acid, the chloride ion $Cl^-(aq)$. This is because the chloride anion is unchanged in the reaction and is present at the beginning and at the end of the reaction.

Figure 4.36
Svante Arrhenius (1859–1927), a Swedish chemist who was awarded the Nobel Prize in chemistry in 1903 for his work on ionic solutions.

Table 4.11 Some acids of major industrial importance.

Acid	Formula
sulfuric	H_2SO_4
phosphoric	H_3PO_4
nitric	HNO_3
hydrochloric	HCl

■   What would you expect to happen if the metals magnesium and tin were treated with hydrochloric acid? Write appropriate equations to describe the reactions.

■   As metals, magnesium and tin would be expected to react with hydrogen cations in the acid to form hydrogen and the metal cation:

$$Mg(s) + 2H^+(aq) = Mg^{2+}(aq) + H_2(g)$$
$$Sn(s) + 2H^+(aq) = Sn^{2+}(aq) + H_2(g)$$

There is another observation that could be made if you carried out reactions of metals with acids but you may be able to guess what it is. Look at the activity series for the metals in Table 4.10. This activity series holds for the reaction of metals with hydrogen cations. Magnesium lies above zinc in the series which in turn is above tin. The reaction of magnesium with hydrogen cations is more rapid than is the corresponding reaction for zinc or for tin. However, it is not always the case that the speed of a reaction is a guide to relative reactivity, although for the reactions of metals with acids it does work reasonably well.

Arrhenius also defined a **base** as a compound that gives hydroxide anions $OH^-(aq)$ in water. Sodium hydroxide $NaOH$ is a white solid comprising the ions Na^+ and OH^-. It dissolves in water to give a basic solution, a solution containing $OH^-(aq)$ ions:

$$NaOH(s) = Na^+(aq) + OH^-(aq)$$

So what happens when an acid and a base are mixed?

Mixing solutions of hydrochloric acid and sodium hydroxide might be expected to give a mixture of the four ions in solution: $H^+(aq)$, $Cl^-(aq)$, $Na^+(aq)$ and $OH^-(aq)$. However, there is a very rapid reaction between the hydroxide and hydrogen ions to give water molecules, a reaction that can be represented by the equation

$$H^+(aq) + OH^-(aq) = H_2O(l)$$

This is **neutralization**, a process in which hydrogen cations and hydroxide anions react together to form water molecules.

Our mixture contained not just the hydrogen and hydroxide ions but sodium and chloride ions, too. What has happened to them?

The simple answer is nothing. The sodium and chloride ions were there at the moment of mixing the two solutions and are there after the hydrogen and hydroxide ions have reacted. That is why they are not represented in the equation. A solution containing equal numbers of sodium and chloride ions is simply a solution of sodium chloride in water.

One could write the overall equation as

$$H^+(aq) + Cl^-(aq) + Na^+(aq) + OH^-(aq) = H_2O(l) + Na^+(aq) + Cl^-(aq)$$

but why bother? The whole point of a chemical equation is to show just what is happening, which substances are reacting and being changed into new substances. The sodium and chloride ions are not reacting so they do not feature in the equation.

■ A solution of nitric acid contains the ions H$^+$(aq) and NO$_3^-$(aq) and a solution of the base potassium hydroxide comprises the ions K$^+$(aq) and OH$^-$(aq). Try to write an equation for the reaction that occurs when these two solutions are mixed.

■ We know that hydrogen and hydroxide ions react to give water molecules, so the equation for the reaction is

$$H^+(aq) + OH^-(aq) = H_2O(l)$$

The potassium ions and nitrate ions are present at the moment of mixing the two solutions and are there, unreacted, in solution at the end.

■ What is striking about the two reactions between hydrochloric acid and sodium hydroxide and between nitric acid and potassium hydroxide?

■ The two reactions are the same. They can both be represented by the equation

$$H^+(aq) + OH^-(aq) = H_2O(l)$$

These are **net ionic equations**. They must balance as normal chemical equations in terms of the relative numbers of atoms on each side of the equation and they must also balance in terms of charge.

Acids and bases are easily recognized by the effect that they have on the colour of some natural dyes known as indicators. The colours of these dyes indicate the concentration of the ions H$^+$ or OH$^-$ in solution. One such dye is the colouring material in red cabbage (Figure 4.37).

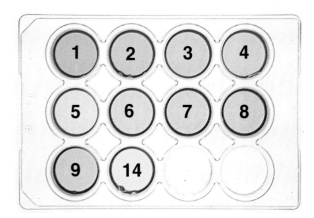

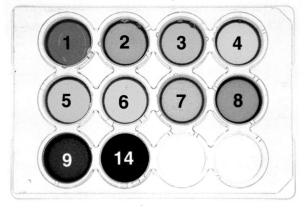

Figure 4.37
Colours of red cabbage juice in solutions of different hydrogen cation concentrations; numbers represent pH values of solutions in cells. The relationship between pH and H$^+$ ion concentration is illustrated in Figure 4.39 (p.94).

Figure 4.38
Colour range of universal indicator. The H$^+$ ion concentration of the solutions is shown by the numbers, which represent pH values.

A more common indicator in the chemistry laboratory is known as 'universal indicator' and comprises a mixture of dyes. The colours shown in Figure 4.38 span a range of H$^+$ ion concentrations differing by a factor of 10^{14}. The H$^+$ ion concentration in cell pH1 is 100 000 000 000 000 times greater than it is in cell pH 14.

We have seen that hydrogen ions $H^+(aq)$ and hydroxide ions $OH^-(aq)$ react together to give molecules of water. At this stage, we need to introduce an idea that is perfectly general and applies to all chemical reactions. When hydrogen cations and hydroxide anions react together, almost all the ions are converted to water molecules, only a tiny concentration of hydrogen cations and hydroxide anions remaining.

To give you some idea of the numbers involved, imagine an experiment which involves mixing an aqueous solution containing one mole of hydrochloric acid with a solution containing one mole of sodium hydroxide. Let's assume that the volume of the resultant solution is exactly one litre.

The reaction is the neutralization

$$H^+(aq) + OH^-(aq) = H_2O(l)$$

However, when the reaction is over, there will still be 1.0×10^{-7} mol of H^+ ions and 1.0×10^{-7} mol of OH^- ions remaining. It does not matter which combination of the acids and which bases you use, there will always be these amounts of hydrogen and hydroxide ions remaining in solution. Even in pure water, there is a quantity of hydrogen and hydroxide ions present.

■   What are the concentrations of hydrogen and hydroxide ions at the end of the experiment?

■   Remember that the final volume of the solution is one litre. The concentration of hydrogen cations is 1.0×10^{-7} mol l^{-1} and the concentration of hydroxide anions is also 1.0×10^{-7} mol l^{-1}.

At 25 °C, the concentrations of the hydrogen and hydroxide ions in water when multiplied together is always equal to 1.0×10^{-14} mol^2 l^{-2}. This seems to be a rather strange observation but let us see just what it means. We shall use the **square bracket** shorthand to write concentrations as $[H^+(aq)]$ and $[OH^-(aq)]$ for the hydrogen cation and the hydroxide anion respectively.

At the end of our experiment, we had

$$[H^+(aq)] = 1.0 \times 10^{-7} \text{ mol l}^{-1}$$

$$[OH^-(aq)] = 1.0 \times 10^{-7} \text{ mol l}^{-1}$$

So $[H^+(aq)] \times [OH^-(aq)] = (1.0 \times 10^{-7} \text{ mol l}^{-1}) \times (1.0 \times 10^{-7} \text{ mol l}^{-1}) = 1.0 \times 10^{-14}$ mol^2 l^{-2}.

(Remember from Book 1 that to multiply such numbers, all you have to do is add the powers to which 10 is raised. For example, $10^2 \times 10^3 = 10^{(2+3)} = 10^5$.)

> **Question 22** What is the concentration of the hydroxide anion in aqueous solutions that contain the following concentrations of hydrogen cations?
>
> (a) 1.0×10^{-14} mol l^{-1} (c) 1.0×10^{-7} mol l^{-1}
>
> (b) 1.0×10^{-4} mol l^{-1} (d) 1.0 mol l^{-1}

The hydrogen cation concentration and hydroxide anion concentration in water are not independent, the higher the hydrogen cation concentration, the lower the hydroxide anion concentration. For example, the cell with pH 14 in Figure 4.38 has a very low concentration of hydrogen cations

(1×10^{-14} mol l^{-1}) but it does have a high concentration of hydroxide anions (1.0 mol l^{-1}). Solutions with a high concentration of hydroxide anions are termed **basic** and solutions with a high concentration of hydrogen cations are **acidic**.

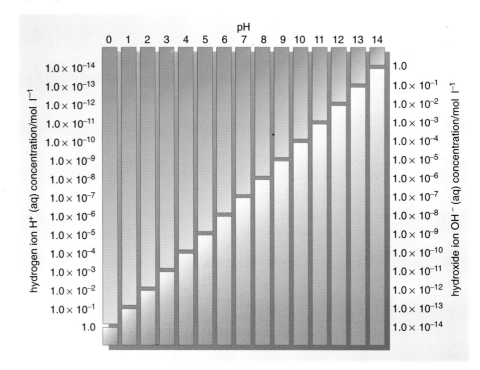

Figure 4.39
Representation of the concentrations of hydrogen and hydroxide ions in different solutions.

In Figure 4.39, in addition to concentration values, there are labels for each of the solutions that indicate pH. pH is a way of denoting H$^+$ ion concentration that avoids the need to use exponents. All you have to do to obtain the pH value is take the exponent in the concentration value and change the sign. When the concentration is written as 1×10^{-n}, the pH value is equal to n. The higher the concentration of hydrogen cations in a solution, the smaller the pH value.

> **Question 23** Express the hydrogen cation concentrations in Question 22(a)–(d) as pH values.

Increasingly, the term pH is creeping into everyday language. It is possible to buy hair shampoo that is 'pH balanced' and there are test kits to measure the pH of garden soil. The acidity level of soil is critical to the successful growth of plants, with some plants only being tolerant to a limited pH range. You will see that pH is an important idea in understanding aspects of nutrition that are explored in Book 3.

> **Experiment 2.9 Acidity of household chemicals**

4.5.2 Oxidation and reduction

The term **oxidation** originally had the simple meaning of combining with oxygen. The definition of **reduction** was that combination with hydrogen had occurred. Today, the meaning is rather wider.

Solid copper oxide CuO is an ionic compound comprising a lattice of the ions Cu^{2+} and O^{2-}. The reaction between elemental copper and oxygen molecules to form the oxide can be classically described as an oxidation:

$$2Cu(s) + O_2(g) = 2CuO(s)$$

Copper has combined with oxygen to form the oxide and so copper is said to be oxidized. Let's look at the reaction a little more closely. The reaction product is the ionic compound, copper oxide. It is helpful to see what is going on if we break away from our convention and write the formula of copper oxide as $Cu^{2+}O^{2-}$ to emphasize its ionic nature; this is just a device to help visualize what is happening. The equation becomes

$$2Cu(s) + O_2(g) = 2Cu^{2+}O^{2-}(s)$$

In the reaction, elemental copper may be viewed as losing electrons to form the Cu^{2+} ion. In this reaction with oxygen, there is a *loss of electrons* by copper atoms. Copper has been *oxidized* in this reaction.

Oxygen has gone from the elemental form to the oxide ion O^{2-}. There has been a *gain* of two electrons. Whenever there is an oxidation, there is always a parallel reduction.

- Copper has been oxidized and oxygen reduced.
- Loss of electrons represents an oxidation.
- Gain of electrons represents a reduction.

The importance of the idea of oxidation and reduction can be seen by looking at the activity series of metals in Table 4.10. The higher metals in the table are those that are more easily oxidized, i.e. lose electrons most easily. The addition of a metal higher in the table to a solution of the cations of a metal lower in the table will result in the reduction of the lower metal ion. For example, if manganese metal is added to a solution containing $Cu^{2+}(aq)$ ions, then the copper cations will be reduced to the metal according to the equation

$$Mn(s) + Cu^{2+}(aq) = Mn^{2+}(aq) + Cu(s)$$

The reverse reaction in which elemental copper is added to a solution of $Mn^{2+}(aq)$ ions does *not* take place. Manganese is more easily oxidized than copper.

Question 24 Use Table 4.10 to predict the outcome of the following:

(a) The addition of Cu to an aqueous solution containing $Ag^+(aq)$.

(b) The addition of Sn to an aqueous solution containing $Cu^{2+}(aq)$.

(c) The addition of Ag to an aqueous solution containing $Cu^{2+}(aq)$.

Table 4.10 can help us rationalize the ease with which metals are obtained from their ores. Copper is towards the bottom of the table, indicating that the metal is not very reactive. It is sometimes found in nature as the element and can be produced from oxides and sulfides by heating with carbon at the relatively modest temperature of 1 000 °C. Iron is a tougher proposition. It is never found naturally as the metal, except in meteorites. The blast furnace process which produces iron from its oxides and carbon needs a temperature of over 1 500 °C. This is a reflection of the tenacity with which iron as the oxide hangs on to oxygen and the reactivity of the metal is indicated by its mid-table position. Higher in the table are metals like aluminium, sodium and potassium.

■   Under what conditions might you guess that aluminium metal could be extracted from its oxide, Al_2O_3?

■   Given that aluminium is higher in Table 4.10 than iron, the metal is likely to be more difficult to obtain from its oxide. It would be reasonable to guess that heating the oxide with carbon at a temperature greater than 1 500 °C would be required.

It is possible for aluminium oxide to be reduced by carbon to aluminium but a temperature of over 2 300 °C is needed. An 'aluminium blast furnace' could not be built and operated economically at this temperature. There would be problems with finding heat-resistant materials with which to line the furnace and the additional expense of operating the plant at such high temperatures. A clue to how aluminium may obtained is seen if we regard aluminium oxide as comprising the ions Al^{3+} and O^{2-}. What is needed is a way of supplying electrons to the aluminium cations.

An electric current is simply electrons moving through a conductor, so why not use these electrons to reduce the aluminium cations? The reaction is carried out in a vessel made from steel with a carbon lining (Figure 4.40). The vessel is known as an electrolytic cell and contains molten aluminium oxide at a temperature of 1 000 °C. The melting temperature of the pure oxide is 2 045 °C. The addition of a naturally occurring fluoride of aluminium, cryolite, to aluminium oxide lowers the melting temperature of the mix. The circuit is completed with carbon rods dipped into the melt. Electrons are fed into the cell via the carbon lining to bring about the reduction of Al^{3+} to the element.

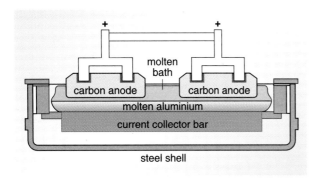

Figure 4.40
A representation of an electrolytic cell for the production of aluminium.

Industrial electrolytic cells are about 10 m in length and 3 m wide and a typical plant will have up to 1 000 cells. Huge quantities of electricity are consumed and it is no surprise that the major aluminium plants are located close to sources of cheap electricity. For example, Jamaica has large deposits of aluminium oxide (the mineral bauxite) but has no aluminium production plants. Much of the country's bauxite goes to Canada where cheap electricity from hydroelectric schemes is exploited. The metals above aluminium in Table 4.10 are extracted by similar electrolytic methods.

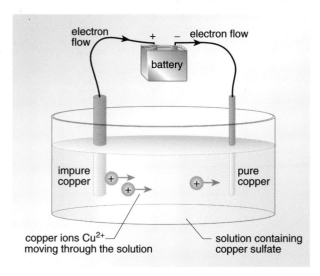

Figure 4.41
Schematic diagram for the electrolytic purification of copper.

Electrolysis can be used to purify metals as well as to win them from their ores. You saw that the output of the extraction of copper was a material known as blister copper. Many of the uses of copper depend on its excellent qualities as an electrical conductor but only when it is of high purity. Electrical conductivity is much diminished by the presence of impurities in the metal, so there is a need for high purity copper. This is obtained by the electrolysis of blister copper.

An electrolytic cell is set up with one electrode being an ingot of impure blister copper and the other a thin rod of the pure metal. These electrodes are dipped into an aqueous solution containing copper ions Cu^{2+}(aq) (this is usually a solution of copper sulfate in water). The circuit is completed by connection to an electricity supply (Figure 4.41).

Electrons are generated at the left electrode which is the ingot of impure blister copper. These electrons are produced when the copper atoms are oxidized. The resultant copper cations go into solution:

$$Cu(s,blister) = Cu^{2+}(aq) + 2e^-$$

The electrons are driven through the external circuit to the right electrode which is in contact with copper cations in solution. At this electrode, reduction occurs. Copper cations in solution are reduced to the metal which is deposited on the electrode:

$$Cu^{2+}(aq) + 2e^- = Cu(s,pure)$$

Only copper atoms are deposited and so the build up of pure copper on the cathode increases. The blister copper electrode (which acts as the anode) is constantly losing copper. Copper atoms of the anode lose electrons and enter the solution as Cu^{2+}(aq) ions replacing those that are removed from solution at the cathode. The overall result is the transfer of pure copper from the blister copper electrode to the pure copper electrode. Impurities drop to the cell floor as impure copper disappears.

This technique can also be used to copper-plate objects. The orchid featured in Figure 4.42 was once living. It was sprayed with a conducting material and then put in a cell similar to that shown in Figure 4.41 in place of the pure copper cathode. Copper was deposited over the orchid and the resulting 'copper orchid' was decorated and sold as a brooch.

Figure 4.42
Painted 'copper' orchid.

Box 4.6 Metal ions and the Periodic Table

Most of the elements in the Periodic Table are metals and these are shown in Figure 4.43. The non-metals are confined to the right of the Periodic Table.

As you have seen, metals are distinguished chemically from the non-metals in that the metals tend to form cations rather than anions. Metals are more easily oxidized than are non-metals. But how do we know what is the appropriate charge for a particular metal cation?

The question is really how many electrons does the metal lose in becoming a cation or what is the value of n in the equation where the metal is represented by the symbol M:

$$M(s) = M^{n+}(aq) + ne^-$$

Table 4.12 gives the charges on a range of common metal and non-metal ions. The ions have been arranged in the table in groups corresponding to the Groups of the Periodic Table into which the elements fall. Look carefully at Table 4.12 and see if there are any generalizations to be drawn.

You may have made some of the following observations:

Table 4.12 Charges on some common cations and anions.

Element	Ion	Charge
Group 1:		
lithium	Li^+	+1
sodium	Na^+	+1
potassium	K^+	+1
caesium	Cs^+	+1
Group 2:		
magnesium	Mg^{2+}	+2
calcium	Ca^{2+}	+2
barium	Ba^{2+}	+2
Group 3:		
aluminium	Al^{3+}	+3
Group 5		
nitrogen	N^{3-}	−3
phosphorus	P^{3-}	−3
Group 6:		
oxygen	O^{2-}	−2
sulfur	S^{2-}	−2
Group 7:		
chlorine	Cl^-	−1
bromine	Br^-	−1
iodine	I^-	−1

The charge on the cations is the same as the Group number of the element from which they are formed. For example, aluminium is in Group 3 and the charge on the cation is +3, Al^{3+}. Sodium is in Group 1 and gives the ion Na^+ with a charge of +1. (Note that lead and tin are exceptions to this rule; they are in Group 4 but do form the ions Pb^{2+} and Sn^{2+}.)

Charges on non-metal ions are negative and are related to but not equal to the Group number. The relationship is that the charges are equal to the Group number minus 8 (although there are some exceptions). Chlorine is in Group 7 and the charge on the chloride ion, Cl^-, is $(7 − 8) = −1$. Sulfur is in Group 6 and the charge on the sulfide ion, S^{2-}, is $(6 − 8) = −2$.

These observations are related to the discussion of electronic configuration of the elements in Book 1.

Question 25 Use Table 4.10 to predict what would happen if:

(a) Fe is added to an aqueous solution containing Sn^{2+} ions.

(b) Al is added to an aqueous solution containing Fe^{2+} ions.

(c) Zn is added to an aqueous solution containing Mg^{2+} ions.

You should have concluded that there would be a displacement reaction in (a) and (b). However, if you actually carried out the experiment in (b), you would find that aluminium would not displace iron from a solution containing iron cations. So what has gone wrong with our activity series?

Figure 4.43 (opposite) The Periodic Table. Pale yellow boxes represent the metallic elements, non-metals are in the green boxes and semi-metals (with properties between metals and non-metals) are shown in orange.

Group →	1	2	3	4	5	6	7	8
	1 **H** 1.01 hydrogen							2 **He** 4.00 helium
	3 **Li** 6.94 lithium	4 **Be** 9.01 beryllium	5 **B** 10.8 boron	6 **C** 12.0 carbon	7 **N** 14.0 nitrogen	8 **O** 16.0 oxygen	9 **F** 19.0 fluorine	10 **Ne** 20.2 neon
	11 **Na** 23.0 sodium	12 **Mg** 24.3 magnesium	13 **Al** 27.0 aluminium	14 **Si** 28.1 silicon	15 **P** 31.0 phosphorus	16 **S** 32.1 sulfur	17 **Cl** 35.5 chlorine	18 **Ar** 39.9 argon

Period 4:
19 **K** 39.1 potassium — 20 **Ca** 40.1 calcium — 21 **Sc** 45.0 scandium — 22 **Ti** 47.9 titanium — 23 **V** 50.9 vanadium — 24 **Cr** 52.0 chromium — 25 **Mn** 54.9 manganese — 26 **Fe** 55.8 iron — 27 **Co** 58.9 cobalt — 28 **Ni** 58.7 nickel — 29 **Cu** 63.5 copper — 30 **Zn** 65.4 zinc — 31 **Ga** 69.7 gallium — 32 **Ge** 72.6 germanium — 33 **As** 74.9 arsenic — 34 **Se** 79.0 selenium — 35 **Br** 79.9 bromine — 36 **Kr** 83.8 krypton

Period 5:
37 **Rb** 85.5 rubidium — 38 **Sr** 87.6 strontium — 39 **Y** 88.9 yttrium — 40 **Zr** 91.2 zirconium — 41 **Nb** 92.9 niobium — 42 **Mo** 95.9 molybdenum — 43 **Tc** 98.9 technetium — 44 **Ru** 101 ruthenium — 45 **Rh** 103 rhodium — 46 **Pd** 106 palladium — 47 **Ag** 108 silver — 48 **Cd** 112 cadmium — 49 **In** 115 indium — 50 **Sn** 119 tin — 51 **Sb** 122 antimony — 52 **Te** 128 tellurium — 53 **I** 127 iodine — 54 **Xe** 131 xenon

Period 6:
55 **Cs** 133 caesium — 56 **Ba** 137 barium — 71 **Lu** 175 lutetium — 72 **Hf** 178 hafnium — 73 **Ta** 181 tantalum — 74 **W** 184 tungsten — 75 **Re** 186 rhenium — 76 **Os** 190 osmium — 77 **Ir** 192 iridium — 78 **Pt** 195 platinum — 79 **Au** 197 gold — 80 **Hg** 201 mercury — 81 **Tl** 204 thallium — 82 **Pb** 207 lead — 83 **Bi** 209 bismuth — 84 **Po** 209 polonium — 85 **At** 210 astatine — 86 **Rn** 222 radon

Period 7:
87 **Fr** 223 francium — 88 **Ra** 226 radium — 103 **Lr** 262 lawrencium — 104 — 105 — 106 — 107 — 108 — 109

Lanthanides:
57 **La** 139 lanthanum — 58 **Ce** 140 cerium — 59 **Pr** 141 praseodymium — 60 **Nd** 144 neodymium — 61 **Pm** 145 promethium — 62 **Sm** 150 samarium — 63 **Eu** 152 europium — 64 **Gd** 157 gadolinium — 65 **Tb** 159 terbium — 66 **Dy** 163 dysprosium — 67 **Ho** 165 holmium — 68 **Er** 167 erbium — 69 **Tm** 169 thulium — 70 **Yb** 173 ytterbium

Actinides:
89 **Ac** 227 actinium — 90 **Th** 232 thorium — 91 **Pa** 231 protactinium — 92 **U** 238 uranium — 93 **Np** 237 neptunium — 94 **Pu** 244 plutonium — 95 **Am** 243 americium — 96 **Cm** 247 curium — 97 **Bk** 247 berkelium — 98 **Cf** 251 californium — 99 **Es** 254 einsteinium — 100 **Fm** 257 fermium — 101 **Md** 258 mendelevium — 102 **No** 259 nobelium

4.5.3 Corrosion of metals

According to Table 4.10, aluminium should react with oxygen more readily than iron, which is easily oxidized in the presence of water and oxygen. How is it then that aluminium can be used for aeroplanes which are in constant contact with water and oxygen?

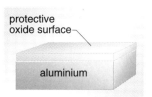

Figure 4.44
Aluminium protected by a layer of oxide.

In reality, aluminium does react with oxygen very rapidly and forms aluminium oxide, Al_2O_3. The oxide has the ability to adhere to the surface of the underlying metal with great tenacity (Figure 4.44). This thin but robust layer of oxide then protects the underlying metal from further oxidation (Figure 4.45).

If the oxide becomes scratched and aluminium metal exposed, more oxide is formed and the protected layer renewed. The reason that no reaction is observed in part (b) of Question 25 is that the oxide film protects the aluminium and prevents reaction with Fe^{2+}(aq) ions. Even if aluminium metal is cleaned with emery paper, a protective oxide layer would form before the metal could be put into the solution containing Fe^{2+}(aq) ions.

The situation with iron is very similar in the sense that iron oxidizes and forms the familiar red–orange rust. However, rust takes up a greater volume than the iron from which it is formed and consequently splits and cracks. It is also porous to water and oxygen so that more rust can be formed from the underlying metal and eventually the oxidation process is complete.

Remember that you saw earlier in Experiment 2.7 that moisture, in addition to oxygen, was also necessary for the rusting of iron. Rust is not the simple oxide Fe_2O_3 but is a compound that actually contains water molecules. It can be represented by $Fe_2O_3.nH_2O$ where n can vary according to the conditions under which the rust forms. The water molecules are incorporated into spaces within the crystal structure of iron oxide.

Figure 4.45
Corrosion-resistant aluminium body of a 1950s Land Rover.

▨   Write a balanced chemical equation to represent the formation of rust on iron.

■ $4Fe(s) + 3O_2(g) + 2nH_2O(l) = 2Fe_2O_3.nH_2O(s)$

By using n, it is possible to write a balanced equation even though we do not know precisely what the formula is for rust.

Iron (and steel) is not weather-resistant yet it does find widespread use because of its relative cheapness and its mechanical properties. Try to think of ways in which steel can be prevented from rusting.

Figure 4.46
Rust spreading under paint on the steel body of a car.

Perhaps the most obvious one is to protect the metal surface with a coating of something that does not rust. The most widely used coating is paint and it is very effective provided that the paint does not become damaged. If this happens, or the metal surface is not completely free of rust when it is painted, then further rusting can occur (Figure 4.46).

The situation is made worse because rusting can spread under the paintwork and is often at an advanced stage when signs of blistered paint become apparent. Where a hard paint surface is not required (such as the inside of car body panels), waxes and paints that do not set completely are

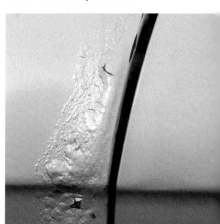

useful. If such a surface suffers a small scratch, the paint will flow a little and effectively self-heal the scratch.

Another coating that protects steel is to use a metal that is further down the activity series, a metal that does not oxidize in the air. The ideal metals are gold silver and copper which are towards the bottom of the activity series in Table 4.10. However, as you can see from Table 4.1, the cost of these metals precludes them except for small-scale specialized uses.

Galvanized buckets and dustbins that used to be commonplace did not rust even though they were made from steel. The metal surface of these containers appears to have irregular shiny crystals (Figure 4.47).

Figure 4.47
Galvanized steel dustbin.

Galvanizing is the process by which a metal surface is given a coating of zinc. Zinc melts at 420 °C, and objects to be galvanized are chemically cleaned and then dipped into a bath of molten zinc (Figure 4.48). The process of galvanizing does seem to work; it is more expensive than conventional paint protection but is finding increasing use in the motor industry.

Where does zinc come in our activity series? We have argued that to coat steel with a metal that does not oxidize in air makes sense, so it seems odd to use a metal that oxidizes even more readily than does iron.

Figure 4.48
Hot-dip galvanizing.

Zinc, like aluminium, oxidizes easily. It also has an oxide that clings well to the metal and is not porous to water or oxygen. So the surface of the steel is protected by zinc metal overlain with zinc oxide. But there is a further advantage to using zinc. If the zinc coating is damaged and iron exposed, one might expect the iron to rust just as if a protective paint surface had been damaged. However, as zinc is more easily oxidized than iron, zinc is oxidized in preference and the iron remains in the metallic form (Figure 4.49).

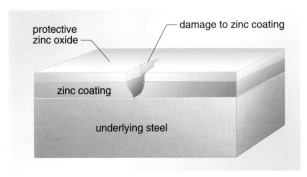

Figure 4.49
Representation of the protection of a galvanized surface. Electrons produced when zinc is oxidized to Zn^{2+} inhibit further oxidation of the underlying iron to the metal cations.

A related protective method is used for larger steel structures such as buildings, ships and underwater/underground pipelines. For such large structures, it would not be practical to provide a covering of zinc. What is done is to attach a block of an easily oxidized metal (magnesium is often used) to the structure by a steel cable (Figure 4.50).

Magnesium preferentially oxidizes to form the Mg^{2+} ion:

$$Mg(s) = Mg^{2+}(aq) + 2e^-$$

The electrons produced can flow along the steel cable and give the oil rig a small negative charge. This makes it difficult for iron in the structure to oxidize and form positively charged ions as this would add even more electrons to what is an already negatively charged structure and the steel rig remains intact. In time, the block of magnesium is eaten away and has to be replaced. This is much easier and cheaper than carrying out repairs on the structure of the rig.

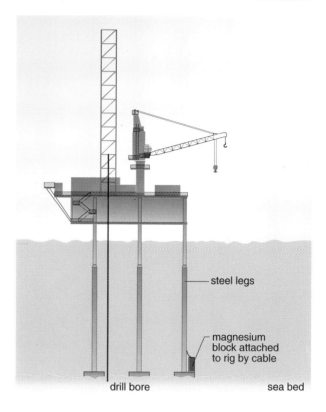

4.6 Modifying metals

In the previous Section, we saw that the resistance to corrosion of a metal could be changed by providing a surface coating of paint or other metal, but there is another way of changing the properties of the metallic elements that has been known for thousands of years. Bronze is an alloy of copper and tin, and like copper, its major component, it is resistant to corrosion. Bronze is also much harder than is copper and it can be used in engineering. It finds particular use in the manufacture of propellers for ships where corrosion resistance is paramount. Here, there is a further advantage: copper is toxic to molluscs and is effective in inhibiting the growth of barnacles on the propeller which would reduce its operating efficiency (Figure 4.51).

Figure 4.50
Protection of the steel in an oil rig platform by using a block of magnesium.

An alloy is different from a chemical compound which has a fixed ratio (and usually a very simple one) of one element to another. We have seen that the relative proportions of copper and tin in bronze can be varied over a wide range. For any ratio, atoms of the two metals are evenly distributed through the structure and the properties of the overall structure are very dependent on that ratio. The hardness, strength and corrosion resistance of metals can often be 'tuned' by carefully adjusting the proportions of elements in the alloy.

Figure 4.51
A ship's propeller cast in bronze.

Steel is an alloy of carbon and iron where small changes in the proportion of carbon can have large changes in the hardness of steel. The addition of other metals to iron can improve corrosion resistance. Stainless steel contains up to 12% chromium by mass often with some nickel. Rusting is prevented in part by a protective layer of chromium oxide. If you look at Table 4.1, you will see that chromium is an expensive metal so stainless steel will never be an economic (although it might be a desirable) alternative to regular carbon steel except for specialized uses. Table 4.13 shows the composition of some of the more common alloys.

Table 4.13 Composition of some common alloys.

Alloy	Composition by mass
bronze	Cu with up to 20% Sn
brass	Cu with up to 35% Zn
stainless steel	Fe with up to 15% Cr and often some Ni
solder	Pb with 40% Sn

It is difficult to predict just what will be the effect of alloying a metal with one or more other elements. There are two main factors to take into account. One is the effect of the different size of the atoms in the alloy and the second is related to the strength of attraction of the atoms of one element to another.

An interesting alloy is formed from copper with about 3% by mass of the element beryllium. Beryllium atoms are rather smaller than those of copper. They seem to effectively 'glue' the copper atoms together to make a hard alloy which has found particular use in fire-sensitive areas such as oil refineries. This alloy will not spark when struck and so lessens the risk of fire.

Another property that is affected by alloying is melting temperature. Solder is an alloy of mainly tin and lead and has a melting temperature of about 180 °C. This is usefully lower than the corresponding temperature for pure tin (232 °C) and lead (328 °C). Intimate contact between the metal atoms in solder and those of the surfaces to be soldered is essential. The metal surface should be cleaned mechanically (with emery paper or steel wool) and then flux applied. Any trace of metal oxide left on the metal surface is reduced to the metal by the flux, improving the quality of the soldered joint.

▪ Using normal methods, it is not possible to solder aluminium but copper is consistently easy to solder. Why should this be the case?

◼ Aluminium is a much more reactive and easily oxidized metal than copper. The surface of aluminium is covered with a thin layer of oxide. It can be removed by rubbing with emery paper but as soon as the metal surface is exposed, the oxide reforms. It is also difficult to reduce aluminium oxide to the metal, and soldering flux is not able to do this. The intervening aluminium oxide surface prevents solder from adhering to the metal. Copper is relatively unreactive and, should there be any oxide on the surface, it will be easily reduced to the metal by hot flux.

One of the distinctive properties of metals is their ability to act as conductors of electricity. Of the non-metals, the allotrope of carbon graphite is the best conductor being, about 10^3 times worse than silver. Silicon comes out about 100 times worse than graphite as an electrical conductor. However, silicon is not such a poor conductor as is sulfur, and silicon is known as a **semiconductor**. This class of materials with conductivities between the metallic conductors and the insulators is vital to the electronics industry. Electronic devices built on the 'silicon chip' have provided the basis for computers, calculators, watches and instrumentation for the past two decades.

A major difference between the metallic conductors and semiconductors is the way the conductivity changes with temperature. This is shown in Figure 4.52 where for clarity we have plotted resistivity rather than conductivity. Resistivity is literally the resistance put up by a material to the passage of an electric current. An insulator will have a very high resistivity and a good conductor a low value.

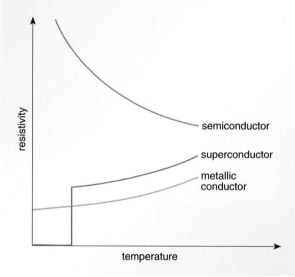

Figure 4.52
Plot of resistivity with temperature.

Metals show an increase in resistivity as the temperature rises – they become poorer conductors. Think back to the model that we used in Section 4.2 for the atomic structure of a metal in which the regular array of metal cations was bathed in a sea of electrons. Flow of an electrical current was effected by movement of this 'sea' of electrons. Close to absolute zero (at −273.15 °C), cations in the metal are essentially motionless and it is relatively easy for the electrons to flow – there is not much resistance. As the temperature rises, the cations begin to vibrate and the vibrations become greater with further temperature rise, thus impeding the flowing electrons. A semiconductor shows a decrease in resistivity as temperature rises. However, at no point do semiconductors become as good a conductor as the metals.

The third line in Figure 4.52 represents a class of materials known as **superconductors**. At very low temperatures, these materials have zero resistivity. Even the best conductors are far from perfect and large currents cause heating, representing an energy loss. Superconductivity would be a most useful property to exploit in electricity transmission because of the tremendous savings in energy and cost that could accrue. Electricity transmission would be more economical, and overhead power lines could be reduced in number and electric motors would be simpler to construct and be more powerful.

Like all promising ideas, there is a downside and in this case it is rather serious. Until 1986, superconductivity had been observed in a number of metals and alloys of metals but only below −250 °C. By any standards, that is cold. Nitrogen liquefies at −196 °C and there are still 54 degrees Celsius to go. This limited the advantages of superconductivity to a few

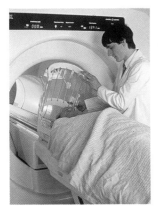

Figure 4.53
Magnetic resonance body scanner.

specialized uses where the expense of the hyper-refrigeration could be justified. One such area is in the creation of large magnetic fields which can be achieved by passing electrical current through wire coils. Superconducting magnets enable large and homogeneous fields to be achieved and this is just what is needed for magnetic resonance body scanners. The bulk of the body scanner shown in Figure 4.53 is a superconducting electromagnet.

The end of the era of metals is not in sight but increasingly we are seeing that new materials are being used in areas traditionally held by metals. One possible replacement for metals is coming from developments in polymer science. Today, about 100 kg of polymers go into a medium-sized car and the amount is increasing year by year. The properties of polymers are being improved by some ingenious structural adjustments at the molecular level. The design of polymers with new properties is the subject of Chapter 5.

Box 4.6 Discovery of high temperature superconductors

When an electric current flows through a wire, there is always some energy loss. The loss depends on a number of factors including the substance from which the wire is made, the thickness of the wire and on the current flowing. Although you would not feel any heat from electricity flowing through the copper cable connecting an electric kettle to a socket on a kitchen wall, there will be a small loss of energy. Electricity flowing through the fine tungsten filament in a light bulb produces sufficient heat to raise the temperature of the filament to over $2\,000\,°C$. In this case, it is of some advantage that there is an energy loss when electricity flows but when large amounts of electricity have to be transmitted over great distances from power stations to consumers, energy loss represents a major cost.

Once electricity is flowing in a superconducting circuit, it will continue to flow without any energy loss. The temperature at which materials become superconductors is termed the transition temperature, T_S. Until 1986, the record for the highest T_S was just $-250\,°C$. In that year, a compound of the oxides of the metals lanthanum and copper to which some barium had been added was found to have a T_S value of $-238\,°C$. This may not seem to be much of a breakthrough but it opened the way to a new class of superconducting materials. Intense research over the next few years produced materials with T_S values as 'high' as $-148\,°C$.

Once the temperature of $-196\,°C$ had been passed, there opened the possibility for commercial exploitation. The gas nitrogen liquefies at $-196\,°C$ and is a relatively cheap coolant. Temperatures lower than this can only be achieved at considerable expense.

Summary of Chapter 4

Metals are better conductors of heat and electricity than are non-metals. They are generally malleable, shiny and lustrous. The difficulty (and therefore cost) of obtaining metals from the Earth depends on concentration and accessibility of the ore and on the ease with which the metal can be extracted from the ore.

Some of the properties of a metal can be explained by a model which comprises metal ions in a 'sea' of electrons. The physical properties of metals, particularly hardness and corrosion resistance, can be affected by the addition of other elements.

Only gold, silver and occasionally copper are found native on Earth. All other metals occur naturally in an oxidized form. Iron is extracted from its oxide by heating the oxide with carbon at high temperature. Aluminium, like other metals high in the activity series, is extracted from its oxide by electrolysis. The activity series represents the ease with which metals can be oxidized. Metals such as sodium, potassium and aluminium are near the top of the series and are easily oxidized. Zinc, iron and tin lie near the middle and the least easily oxidized metals, gold, silver and copper, lie near the bottom of the series. Oxidation is loss of electrons and reduction is gain of electrons.

Problems associated with the rusting of metals may be minimized by coating the surface of the metal with a protective layer such as paint or by galvanizing with zinc, by alloying with another metal or by using a sacrificial metal.

Balanced chemical equations can be used to calculate the relative amounts of reactants and products in a reaction. Chemical equations representing ionic reactions must balance both in terms of the quantity of material and the total charge on each side. An acid produces hydrogen ions (H^+) in solution and a base, hydroxide ions (OH^-). In water, the neutralization reaction is the same for all acids and bases. In such a solution, $[H^+][OH^-] = 1 \times 10^{-14}$ mol^2 l^{-2}.

Question 26 Calculate the amount of carbon required to react with one tonne of the oxide Fe_2O_3 to give metallic iron and carbon dioxide according to the equation

$2Fe_2O_3(s) + 3C(s) = 4Fe(s) + 3CO_2(g)$

Question 27 Balance the following equations:
(a) $Fe(s) + H^+(aq)$ /=/ $Fe^{2+}(aq) + H_2(g)$
(b) $Ca(OH)_2(s) + H^+(aq)$ /=/ $Ca^{2+}(aq) + H_2O(l)$
(c) $CrO_4^{2-}(aq) + H^+(aq)$ /=/ $Cr_2O_7^{2-}(aq) + H_2O(l)$

Question 28 Nitric acid (HNO_3) is in the form of $H^+(aq)$ ions and $NO_3^-(aq)$ ions in solution. Calculate the concentration of hydrogen ions $H^+(aq)$ if a solution of volume 200 cm^3 contains 63.01 g HNO_3.

Question 29 Write a balanced chemical equation for the reaction between an aqueous solution of nitric acid HNO_3 and a sodium hydroxide NaOH solution.

Question 30 Write a balanced chemical equation to show the reaction between a solution of phosphoric acid H_3PO_4, and magnesium hydroxide, $Mg(OH)_2$. Phosphoric acid dissolves in water to give a solution containing hydrogen ions $H^+(aq)$ and phosphate ions PO_4^{3-} (aq), and magnesium hydroxide is a relatively insoluble ionic solid.

Question 31 Distinguish between the terms isotope, allotrope and isomer and give examples of each.

Chapter 5
Polymers

We shall now take a look at some of the materials around us that we collectively call 'plastics', although the more correct term to use is 'polymer'. You should recall this term from the article on fibres in Book 1, which implied that the word embraces many natural materials, such as natural fibres, as well as many modern 'high technology' materials that we tend to belittle by using the term 'plastic'. In this Chapter, we shall examine how natural materials have been modified and synthetic ones developed to meet our continually changing needs. We shall see how materials are designed at the molecular level to influence the properties that we see and sense at the macroscopic level. We shall also look at how synthetic polymers – materials of great importance in commerce, industry and our daily lives – are major consumers of raw materials and lead to large amounts of waste. Their potential for harming the environment can be assessed, and routes to recycling offer possibilities for sustainable development.

The AV sequences entitled 'Organic chemistry' introduce a topic called the functional group approach, which will be of great use to you as study this Chapter and also, particularly, Books 3 and 4. You will find it very helpful to listen to the first part of the sequence now.

5.1 Molecules in everyday life

Look back to Figures 1.6 to 1.9 at the beginning of this Book and note some of the types of material used in daily life throughout the ages. Among them, you will notice many string-like materials or fibres. This appearance at the macroscopic level is often a reflection of the structure at the molecular level, as we shall soon see. However, you may also notice a major difference between the last two pictures, a century apart. During the past 50 years, we have entered what could justifiably be called the age of plastics.

What do body implants, pencil erasers, UPVC window frames, superglue, bullet-proof garments, many fabrics, a television or computer case and a plastic carrier bag have in common with dragonfly wings, paper, wood, flesh and even the genetic material itself, DNA? They are all made up of materials that, at the molecular level, consist of very long molecules – far longer than most of the chemicals you have met so far in this Course. These materials, and many more that we encounter in daily life, are called polymers. A small selection appears in Figure 5.1.

Activity 1 Polymers in the home

Look around your home and make your own list of all the materials you regard as polymers or 'plastics'. Note the appearance and feel of the different materials and write down any special properties of each that make it most suited to its function. By the time you have completed this Chapter, you should be able to explain some of these differences. While at present you may not be able to name all of the polymers you find, near the end of this Chapter we'll give you some clues to their identification through a coding system to be found on many plastic products.

Figure 5.1
Polymers in our daily life.

Many polymeric materials are products of nature and have been extracted and processed chemically for use in everyday products. Others are not natural and are the results of chemical innovations throughout the 20th century. Given the huge range of uses of polymers today, we can hardly imagine what modern life would be like without polymers! For example, the ubiquitous motor car may be a source of pleasure, an essential business tool, a fashion accessory or a modern source of nuisance. Whatever its role, strip away the various plastic components (Figure 5.2) and you will be left with a very bare product that would have difficulty meeting our expectations of safety, efficiency and comfort.

Figure 5.2
Exploded view of a car body showing the various plastic components.

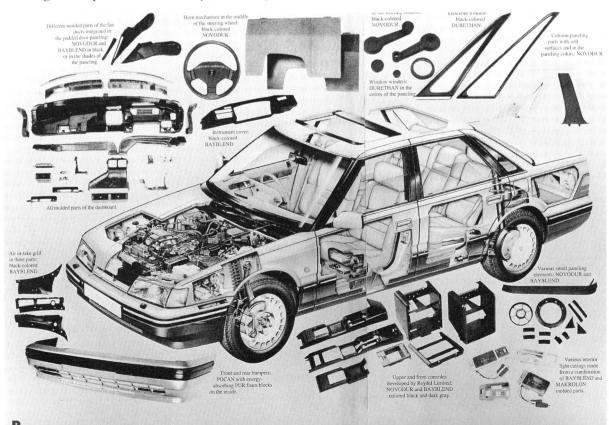

Box 5.1 The perspex beetle

When attacked, the beetle *Abax parallelepipedus* (Figure 5.3) produces a secretion which is mainly a chemical called methyl methacrylate. When this is fired at an attacker, it forms a polymer and immobilizes the enemy. We know this polymer as 'perspex', a product we usually associate with modern chemical technology.

Figure 5.3
The beetle that found a use for Perspex before the chemical industry.

Synthetic polymers are widely used in modern medicine and in particular to reconstruct or restore certain parts of our bodies. Most human tissues consist of very complex polymers, which we are at present unable to replicate. Even so, there are a number of important applications in which lost function can be restored by the use of modern synthetic materials. Heart valves and soluble stitches are examples of polymers serving human health.

Just as we know of the Stone Age, Iron Age and Bronze Age of the past, the sequel to the Industrial Revolution has been the polymer revolution. However, long before modern polymers were invented and before the word 'polymer' was even coined, ancient civilizations used many natural polymers, as Figures 1.6 to 1.9 at the beginning of this Book testified. For food, we have always used animal proteins and starch, which are natural polymers. Our ancestors cooked food on fires of wood, which contains cellulose and is similar in some respects to starch. Polymers were also used for clothing. Cellulose is also found in flax, which people are known to have cultivated long ago; the proteins wool and silk were also used by ancient civilizations. Comparing these polymers with modern-day plastics, adhesives and paints, it is evident that modern materials are totally different in terms of their appearance, texture, hardness, flexibility and so on (Figure 5.4).

Clearly, the polymer revolution has had an enormous impact on our lifestyles, but is not without its environmental problems, which result to a large extent from the ideal properties of modern polymers – their robustness and longevity. In Section 5.7, we'll see how chemistry can offer solutions to such problems.

Figure 5.4
Some modern polymers and their uses.

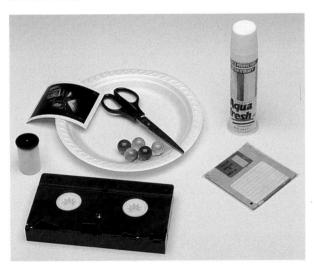

5.2 What is a polymer?

The word **polymer** is a chemical term originally coined by the chemist
J.J. Berzelius in 1827, and it has now entered everyday language. It is
derived from the Greek words *poly* meaning 'many' and *meros* meaning
'parts', and so we can conclude that a polymer is built up by joining
together many small chemical building blocks that are often, but not
exclusively, identical. The chemical building blocks from which the
polymer is made are called **monomers** (meaning one part).

In a crude model, paper-clips could represent our building blocks or
monomers. Imagine that you have collected together lots of paper-clips
and interlinked them to form a long chain. An individual paper-clip
represents one of the monomers, the chain of paper-clips represents a
polymer chain or polymer molecule. Such a paper-clip chain would be
very flexible. It would have no fixed form or arrangement. In this respect,
the chain is a reasonable model of a polymer molecule.

Many monomers are manufactured from crude oil or petroleum, and the
process of joining them together into long chains is called
polymerization. Polymer molecules are rather different from most of the
molecules that you have met previously in this Course – they are much
longer!

How long are these long molecules? Think back to Chapter 7 of Book 1,
and recall how by starting from methane, with one carbon atom, you built
up to a longer molecule such as hexane with a chain of six carbon atoms.
By comparison, polyethene (or polythene as it is more commonly known)
for packaging may have many thousands of carbon atoms joined together
in a chain.

The concept of polymers being very large molecules was introduced by
Hermann Staudinger (1881–1965), who advanced the theory that polymers
were not simply aggregates of small molecules but were large structures
held together by strong chemical (covalent) bonds. You can appreciate the
difference between these concepts by considering how your imaginary
paper-clips would aggregate together weakly or tentatively if you
magnetized them, compared with the strong links formed by interlinking
them. Before polymerization the forces between monomers are only weak,
for example London forces (introduced in Chapter 8 of Book 1), and these
are not strong enough to build up long and stable chains. Strong covalent
bonds are necessary to form long polymer molecules.

Some polymers exhibit properties we don't normally associate with
plastics: they may be as strong as steel, conduct electricity, resist high
temperatures or have other characteristics. Yet they can all be classed
together as polymers because they share the property of being composed
of long-chain molecules derived from monomers. To understand how we
achieve such a wide range of physical properties from polymers, we need
to know something about their molecular structures or how each polymer
is built up from its basic building blocks. From this we shall begin to see
how the characteristics we desire at the 'macroscopic level' (i.e. the
properties we want in terms of strength, appearance, flexibility, etc.)
depend on the characteristics at the molecular level. Polymers are one of

many examples of 'designer chemicals' you will meet in this Course and we'll see how such products have been developed, and also how some have resulted by chance.

We shall begin with the simplest polymer of all, polyethene, and then move on to fibres. The appearance of fibres reflects most clearly the underlying shape of their constituent polymers: long-chain molecules. First, though, we need to finish this Section with something about the names of polymers.

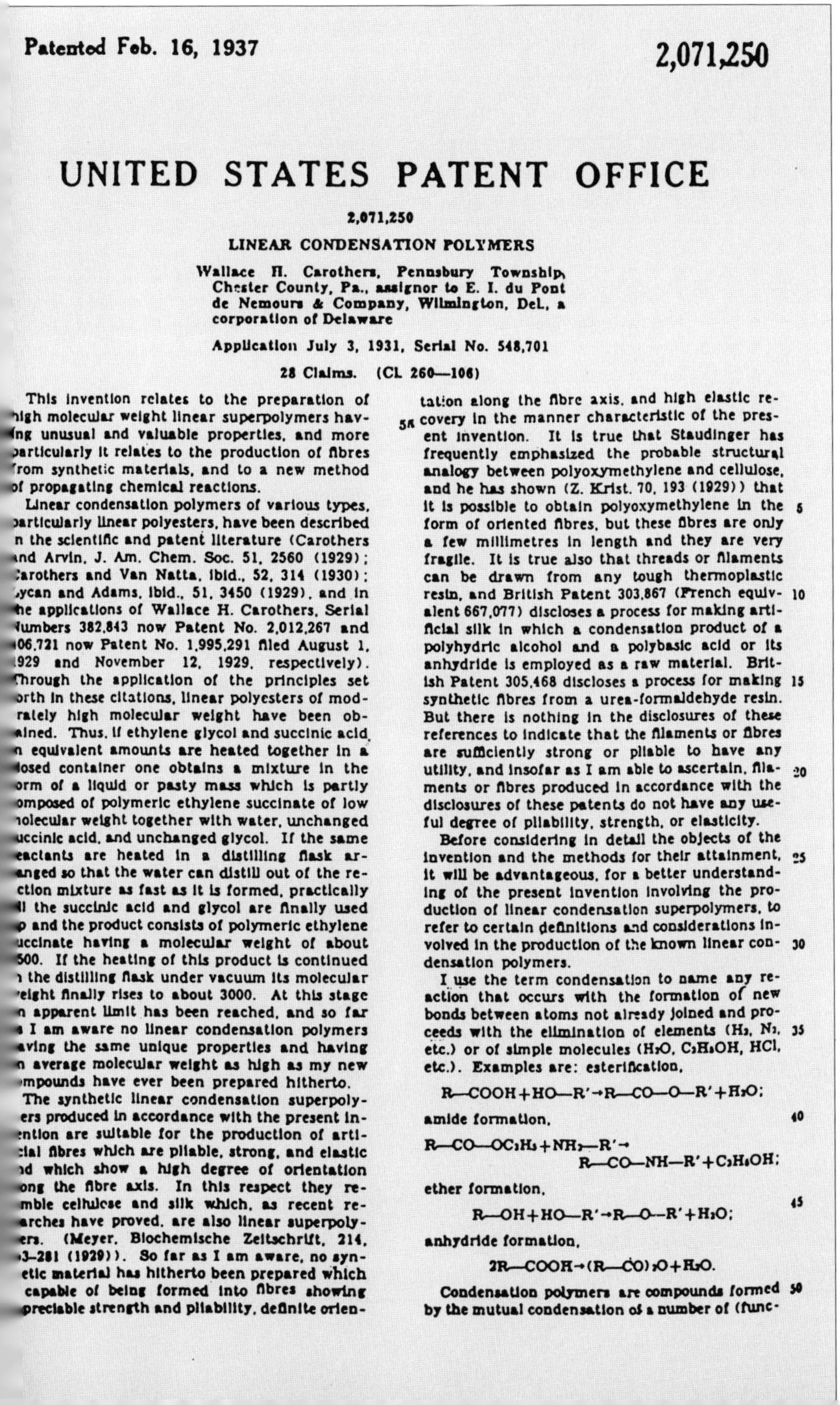

5.2.1 What's its name?

Although modern names of chemicals are derived systematically, the traditional names of numerous chemicals often have strange origins, and many remain in common use. You should recall this from Chapter 7 of Book 1, where the potential confusion in the common names for a drug was illustrated. Similar problems occur with polymers. Let's take the familiar name 'nylon', the first fully synthetic fibre (Figure 5.5). It is said that the first suggestion for a name for the fibre was 'norun', because stockings made from the fibre resisted snagging. This name was rejected by the company and a series of alterations resulted in the word nylon entering our language. Another story attributes the name to a condensation from New York and London!

Many common names of synthetic polymers have been obtained in a more obvious way than that of nylon. Bakelite, for example, is named after Leo Baekeland who first developed it in 1905, but many polymer names have been derived simply by adding the prefix 'poly' to the common name of the basic building block or monomer. For example, polystyrene is the most frequently used name for the polymer of 'styrene'. This approach to naming is used particularly for polymers made up of a single type of monomer.

Figure 5.5
Patent for nylon.

Patented Feb. 16, 1937

2,071,250

UNITED STATES PATENT OFFICE

2,071,250

LINEAR CONDENSATION POLYMERS

Wallace H. Carothers, Pennsbury Township, Chester County, Pa., assignor to E. I. du Pont de Nemours & Company, Wilmington, Del., a corporation of Delaware

Application July 3, 1931, Serial No. 548,701

28 Claims. (Cl. 260—106)

This invention relates to the preparation of high molecular weight linear superpolymers having unusual and valuable properties, and more particularly it relates to the production of fibres from synthetic materials, and to a new method of propagating chemical reactions.

Linear condensation polymers of various types, particularly linear polyesters, have been described in the scientific and patent literature (Carothers and Arvin, J. Am. Chem. Soc. 51, 2560 (1929); Carothers and Van Natta, ibid., 52, 314 (1930); Lycan and Adams, ibid., 51, 3450 (1929), and in the applications of Wallace H. Carothers, Serial Numbers 382,843 now Patent No. 2,012,267 and 406,721 now Patent No. 1,995,291 filed August 1, 1929 and November 12, 1929, respectively). Through the application of the principles set forth in these citations, linear polyesters of moderately high molecular weight have been obtained. Thus, if ethylene glycol and succinic acid in equivalent amounts are heated together in a closed container one obtains a mixture in the form of a liquid or pasty mass which is partly composed of polymeric ethylene succinate of low molecular weight together with water, unchanged succinic acid, and unchanged glycol. If the same reactants are heated in a distilling flask arranged so that the water can distill out of the reaction mixture as fast as it is formed, practically all the succinic acid and glycol are finally used up and the product consists of polymeric ethylene succinate having a molecular weight of about 500. If the heating of this product is continued in the distilling flask under vacuum its molecular weight finally rises to about 3000. At this stage an apparent limit has been reached, and so far as I am aware no linear condensation polymers having the same unique properties and having an average molecular weight as high as my new compounds have ever been prepared hitherto.

The synthetic linear condensation superpolymers produced in accordance with the present invention are suitable for the production of artificial fibres which are pliable, strong, and elastic and which show a high degree of orientation along the fibre axis. In this respect they resemble cellulose and silk which, as recent researches have proved, are also linear superpolymers. (Meyer, Biochemische Zeitschrift, 214, 253–281 (1929)). So far as I am aware, no synthetic material has hitherto been prepared which capable of being formed into fibres showing appreciable strength and pliability, definite orientation along the fibre axis, and high elastic recovery in the manner characteristic of the present invention. It is true that Staudinger has frequently emphasized the probable structural analogy between polyoxymethylene and cellulose, and he has shown (Z. Krist. 70, 193 (1929)) that it is possible to obtain polyoxymethylene in the form of oriented fibres, but these fibres are only a few millimetres in length and they are very fragile. It is true also that threads or filaments can be drawn from any tough thermoplastic resin, and British Patent 303,867 (French equivalent 667,077) discloses a process for making artificial silk in which a condensation product of a polyhydric alcohol and a polybasic acid or its anhydride is employed as a raw material. British Patent 305,468 discloses a process for making synthetic fibres from a urea-formaldehyde resin. But there is nothing in the disclosures of these references to indicate that the filaments or fibres are sufficiently strong or pliable to have any utility, and insofar as I am able to ascertain, filaments or fibres produced in accordance with the disclosures of these patents do not have any useful degree of pliability, strength, or elasticity.

Before considering in detail the objects of the invention and the methods for their attainment, it will be advantageous, for a better understanding of the present invention involving the production of linear condensation superpolymers, to refer to certain definitions and considerations involved in the production of the known linear condensation polymers.

I use the term condensation to name any reaction that occurs with the formation of new bonds between atoms not already joined and proceeds with the elimination of elements (H_2, N_2, etc.) or of simple molecules (H_2O, C_2H_5OH, HCl, etc.). Examples are: esterification,

$$R-COOH + HO-R' \rightarrow R-CO-O-R' + H_2O;$$

amide formation,

$$R-CO-OC_2H_5 + NH_2-R' \rightarrow$$
$$R-CO-NH-R' + C_2H_5OH;$$

ether formation,

$$R-OH + HO-R' \rightarrow R-O-R' + H_2O;$$

anhydride formation,

$$2R-COOH \rightarrow (R-CO)_2O + H_2O.$$

Condensation polymers are compounds formed by the mutual condensation of a number of (func-

---

Box 5.2 Botanical varieties

From my gardening books, I notice two varieties of 'sweet gum' tree – *Liquidambar styraciflua* and *Liquidambar orientalis*. The resin from the latter is called 'storax' and was used by the Egyptians for embalming. In 1839, Eduard Simon found that a clear liquid could be obtained on heating a mixture of storax and water – this liquid contains styrene. Simon also noticed that on further heating, styrene turned into a solid gelatinous mass, a substance that we now know as polystyrene. He regarded this conversion of a liquid into a solid as remarkable, because most liquids turn into a gas on heating, as pointed out in Chapter 6 of Book 1. This example of polystyrene and the earlier one of perspex illustrate that even what we may perceive as modern synthetic polymers may have a natural occurrence.

We are all familiar with the materials polythene and PVC, which is an abbreviation of polyvinyl chloride. Both of these common names of polymers are derived from the names of their respective monomers, ethene and vinyl chloride. Table 5.1 includes these two examples with several others that are named after their monomers. It appears a daunting list, but we shall look at a only few of these polymers, and there is no need to learn all of them.

Table 5.1 Some common polymers named after the appropriate monomers.

Common name for monomer	Common polymer name (abbreviation)	Examples of trade names	Uses
styrene	polystyrene (PS)	Styrofoam, Cellofoam, Styron, EPS	packaging, insulation, adhesives, instrument panels, bottles, glasses, vending machine cups, toys
ethene (or ethylene)	polyethene, polythene or polyethylene (PE)	Durethene, Pearlon, Dylan	packaging films and sheets, toys, moulded food containers, bottles, tubes, pipes, coatings on paper, car components, etc.
propene (or propylene)	polypropylene (PP)	Dynafilm, Propathene	rope, carpet fibre, car parts, packaging
acrylonitrile	polyacrylonitrile (PAN)	Acrilan, Courtelle, Orlon	textile fibres, carpet fibres
vinyl chloride	polyvinyl chloride (PVC)	vinyl, carina, plioflex, Viclan	pipes and tubing, rainware, records, floor covering, wire insulation, window frames, bottles, films and blister packs, adhesives
vinyl acetate	polyvinyl acetate (PVA)		emulsion (latex or water-based) paint, wood adhesive
methyl methacrylate	poly(methylmethacrylate) (PMMA)	Plexiglas, Perspex	car parts (lenses), display signs, baths
chloroprene	polychloroprene (CR)	Neoprene	oil-resistant rubber items
isoprene	polyisoprene (IR, NR)	natural rubber	rubber items

Another family of polymers is classified according to the chemical name for the link that forms when the monomers react to give a long chain. Such links are called connecting groups, as shown schematically in Figure 5.6. You are sure to recognize some of these polymer names, listed in Table 5.2. For example, we can have polyester fibres, and polyurethane, familiar from paints and many foam products used in packaging or cushions. When you next look at a garment label indicating that the fabric contains polyamide, for instance, you will know that it includes a polymer with a connecting group of the 'amide' type, even though, at this stage, we have not described such a connecting group in any detail.

Figure 5.6
Schematic representation of a polymer. Each block, which is derived from the monomer, is joined to its neighbours by two connecting groups to produce a long chain. The connecting groups are shown as couplings. Polymers are classified according to either the name of the original monomer (Table 5.1) or the chemical name for the connecting group (Table 5.2).

Table 5.2 Some common polymers based on connecting group names.

Connecting group name	Polymer family name	Examples of trade names	Uses
amide	polyamide	Antron, Enkalure, Nomex, Nylon	textile fibres, bristles, carpets, tyre cord, fishing line, engineering components
ester	polyester	Dacron, Trevira, Mylar, Scotchpak, Terylene, PET, Crimplene, Melinex	fabrics, magnetic tape for video etc., bottles for carbonated drinks, oven-proof food trays
carbonate	polycarbonate	Lexon, Merlon, Makrolon	unbreakable glazing sheets, safety glasses and helmets, some packaging
urethane	polyurethane	Isofoam, Spandex, Vithane	upholstery padding, packaging, air filters, paints and lacquer

Whereas simple polymers are most commonly named by prefixing the common name of the appropriate monomer with 'poly' (as in Table 5.1), guidelines for systematic naming have been produced. Once again, the monomer is the basis, but this time the systematic chemical names of the monomers are used. With so many chemicals known (the figure of 12 million was mentioned in Book 1), it is important to be able to name any chemical unambiguously, in a similar manner to which some gardeners refer to the systematic names of plants. Table 5.3 gives systematic names of some common polymers you may read about elsewhere. However, as we pointed out in Book 1, you need not remember systematic names in this Course.

Table 5.3 Systematic names for some common polymers.

Common name	Systematic name
polythene, polyethylene	poly(ethene)
polystyrene	poly(phenylethene)
polypropylene	poly(propene)
polyvinyl chloride	poly(chloroethene)
perspex (polymethyl methacrylate)	poly((1-methoxycarbonyl)-1-methylethene)

Once we have named a polymer by reference to the parent monomer or to the connecting group, we can also classify polymers according to certain physical properties. One common classification is to divide polymers into two main types, known as **thermoplastics** and **thermosets**. Some estimates suggest that about 100 million tonnes of synthetic polymers are used each year worldwide, with about 85% being thermoplastics and 15% thermosets.

Thermoplastics, from which the whole plastics industry has derived its name, are polymers that can be softened at high temperatures and transformed into any desired shape. On cooling, they solidify, but can be reshaped on heating again.

The shapes of thermosets or thermosetting plastics are set by the action of high temperature and cannot be reset by further application of heat. You can appreciate why this is an important feature if you consider some uses of thermosets. They include the earliest plastics, such as bakelite, and continue to be used in applications where high durability is needed, such as for electrical items, pan handles and the sheet materials of kitchen worktops.

> **Question 32** (a) Teflon, used to coat non-stick frying pans, and PTFE tape are made from a polymer with the systematic name of poly(tetrafluoroethene). Write down the name of the monomer used to produce Teflon and PTFE. (b) The acrylic fibre, Acrilan, has the systematic name poly(propenenitrile). Use Table 5.1 to suggest the common name for propenenitrile.

5.3 The chain gang

So far, we have dealt with polymers in only a general way, with little reference to details of their molecular structure. Before looking at fibres and three-dimensional aspects, we need to appreciate more of the underlying chemistry, and our easiest route is to start with polyethene, which has a relatively straightforward chemical chain.

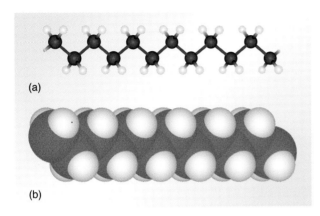

(a)

(b)

Figure 5.7
(a) Ball-and-stick and
(b) space-filling models of a portion of a polyethene molecule.

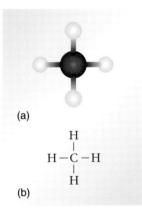

(a)

$$H-C-H$$

with H above and H below the central C.

(b)

Figure 5.8
Ball-and-stick model of methane and an abbreviated two-dimensional representation of the same molecule.

5.3.1 The structure of polyethene

Plastic sheeting, squeezy bottles, washing-up bowls, refrigerator ice trays, milk bottle crates and numerous other items are made from polyethene. But what is polyethene? Figure 5.7 shows a portion of one of its chains. The whole molecule is no more than a string of thousands of CH_2 units, that is, a linear hydrocarbon taken to extremes. In Activity 10 of Book 1, you made a model of hexane with a string of six carbon atoms. You started that exercise by constructing six CH_2 units and then joined them all together so that the carbon atoms were in a row. If you had a more extensive model kit (and infinite patience) you could make thousands of CH_2 units, couple them all together and hence create a model of polyethene. Remember also that you found that hexane was a floppy molecule with many different shapes or conformations. Just think of the number of possible shapes for a polyethene molecule that would be at least one thousand times longer than hexane!

Figure 5.7 shows only one possible shape for part of the polymer chain. Note that, because the ball-and-stick model represents only a small part of a very long molecule, we have left the two bonds at the ends of the chain unconnected. They signify that the structure is incomplete and that it repeats to a considerable length. In reality, there would be further CH_2 units attached to complete carbon's valency of four. You will meet incomplete structures with 'unconnected bonds' at several points in the Course. In this Chapter they denote continuing sequences in structures.

It is very cumbersome to portray polymers as in Figure 5.7. Instead, we can replace the balls in a ball-and-stick model with the appropriate chemical symbol for that atom and replace the sticks with a line representing a chemical bond. To move towards a shorthand method of depicting large structures with this approach, we begin with the simplest hydrocarbon, methane.

As mentioned previously, a carbon atom requires four bonds. As the major constituent of natural gas, methane is a familiar gas. In chemistry, it has the formula CH_4. You should recall the picture of methane as a carbon atom with four links or bonds, each pointing from it and attached to a hydrogen atom (Figure 5.8).

In both Figure 5.8 and Table 5.4, the structure of methane is depicted in two dimensions using chemical symbols instead of coloured balls to depict atoms, and lines to depict the bonds. Note that this representation is not meant to show the three-dimensional shape of methane at all. Adding a CH_2 unit to the molecular formula for methane gives us the molecular formula C_2H_6 and the next member of the hydrocarbon series – ethane (Figure 5.9).

Figure 5.9
Ball-and-stick model of ethane.

■ Without referring to Table 5.4, try to draw a two-dimensional representation of ethane, taking your lead from its ball-and-stick model in Figure 5.9.

■ The exercise is the same as the one that produced the representation of methane in Figure 5.8. You should have replaced the dark grey and white balls in the ball-and-stick model of ethane with the symbols for carbon (C) and hydrogen (H), respectively, and replaced all of the sticks with a line representing a bond. The result should look like the entry for ethane in Table 5.4.

Note in particular that in ethane the two carbon atoms are bonded to each other. We see the beginnings of a linear chain of carbon atoms. Let's continue this building up process. Adding another CH_2 unit gives us propane – often used as a bottled fuel for camping stoves, greenhouse heaters and the like. Similarly, the next member of the series is butane, with four carbon atoms joined together. Both propane and butane can easily be turned into liquids by pressurizing them in cylinders. This behaviour is a feature of the chain getting longer and the London forces between molecules increasing. Hexane, with six carbon atoms and 14 hydrogen atoms (C_6H_{14}), is a liquid fuel at room temperature, as is octane (C_8H_{18}).

When we reach 18 carbon atoms, with their associated hydrogen atoms, the compound is a solid at room temperature. The so-called 'paraffin waxes' contain around 30 carbon atoms, and are used to make candles and to weatherproof fabrics. Hydrocarbons with up to 70 carbon atoms give waxes that have higher melting temperatures and that serve as the base for petroleum jelly and shoe polish. All of these hydrocarbons that differ only in the number of CH_2 units are called alkanes. They occur naturally, but the alkane series can stretch much further as we have seen with polyethene – to 200 000 carbon atoms or more! As it does so, the method of drawing structures as in Table 5.4 rapidly gets unwieldy.

The chains drawn in the Table can be written in a shorter form. Take butane as an example. The first carbon atom is bonded to three hydrogen atoms and hence comprises a CH_3 group. The next carbon atom is bonded to two hydrogen atoms and so forms a CH_2 group. When this is joined to the CH_3 group, the molecule so far can be abbreviated to CH_3-CH_2. This grouping is followed by another CH_2 group and finally a CH_3 group. Butane can thus be written as $CH_3-CH_2-CH_2-CH_3$. Even shorter, we can miss out the carbon-to-carbon bonds and just write $CH_3CH_2CH_2CH_3$. We are now nearly in a position to portray such hydrocarbons, and polyethene in particular, very simply. Given that the linear alkanes contain strings of CH_2 units, rather than write them all out, we can use parentheses and a suffix to represent their number in a chain. In this notation, butane becomes $CH_3(CH_2)_2CH_3$, indicating that two CH_3 groups are linked by two CH_2 groups.

Table 5.4 Some of the shorter-chain hydrocarbons called alkanes.

C atoms in chain	Name	Formula	Structure	Importance as component of:
1	methane	CH_4	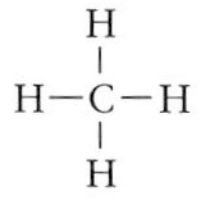 H−C−H (with H above and below)	natural/landfill gas
2	ethane	C_2H_6	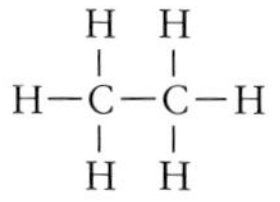 H−C−C−H	
3	propane	C_3H_8	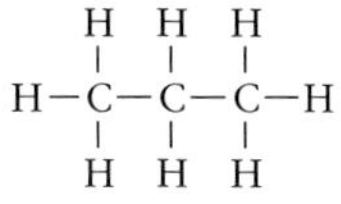 H−C−C−C−H	fuel gas
4	butane	C_4H_{10}	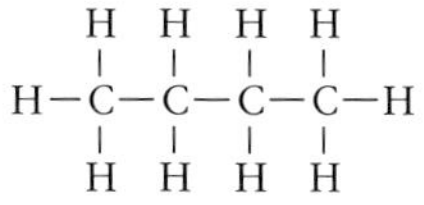 H−C−C−C−C−H	fuel gas
6	hexane	C_6H_{14}	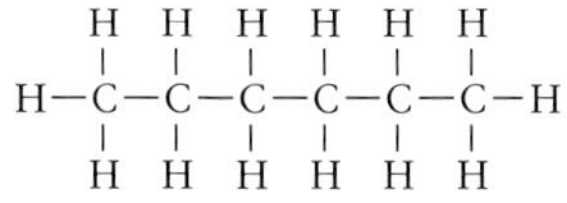 H−C−C−C−C−C−C−H	
8	octane	C_8H_{18}	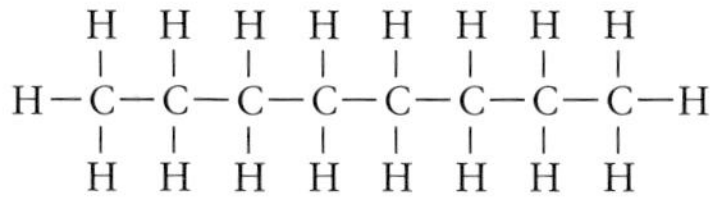 H−C−C−C−C−C−C−C−C−H	petrol

Question 33 Write the shorthand form of octane.

So, how do we write polyethene without using up too much space on a page? If we recognize that polyethene is just a string of CH_2 groups, we can depict it as $CH_3(CH_2)_nCH_3$, where n stands for several thousand. If we were not concerned about the groups that cap the ends of the chain, we could write polyethene as $-(CH_2)_n-$, using the 'unconnected bond' approach. If you are unsure about this new notation, make a model of a portion of a polyethene chain containing six CH_2 groups, as you did on the way to making hexane in Activity 10 of Book 1. You should be able to see that your model is described by $(CH_2)_n$ where $n = 6$. Note, though, that the model allows us to see that the carbon chain has a zig-zag shape whereas the $(CH_2)_n$ notation does not. The penalty we pay for simple depiction is a loss of information on the shapes of molecules.

A stereo picture representing the true arrangement of the polyethene chain is given in Figure 5.10. Use your stereo viewer to compare this Figure with your model.

Figure 5.10
Three-dimensional representation of a polyethene chain (use your stereo viewer to look at this picture).

Question 34 The systematic name of polythene should reveal the name of the monomer from which it is made. Referring to Tables 5.1 and 5.3 if you wish, give the name of the appropriate monomer. It is a molecule that you met in Book 1. Write down what you can recall about its structure and chemistry.

Ethene is the monomer for polyethene. In other words, when ethene molecules react with each other, they form a long chain of CH_2 units that constitute a polyethene molecule. Figure 5.11 shows the ball-and-stick representation of ethene (previously seen as Figure 7.25 of Book 1) and the shorthand version of the same molecule in which the balls are replaced with the chemical symbols for carbon and hydrogen and the sticks (that is, the bonds) are replaced with lines. Note in particular how the double bond between the carbon atoms is portrayed as two parallel lines.

Because of its double bond, ethene is an action-packed molecule. You may recall that there are electrons forming the bonds between atoms, so the region between the carbon atoms is especially rich in electrons, because there are two bonds there. This high density of electrons confers considerable reactivity on ethene. Even so, a sample of ethene is stable in the sense that it does not immediately form polyethene. In fact, something else is needed to start or initiate the reaction to give polyethene. We say that the polymerization needs an **initiator**. An initiator is a very reactive molecule that goes in search of an electron-rich area, such as the double bond in a molecule of ethene, and breaks one of the two bonds, forming a link with one of the carbon atoms in the ethene molecule. The process is illustrated in Figure 5.12. Once it happens, the high reactivity is passed on to the other carbon atom in ethene, which then reacts with another molecule of ethene. Each time an ethene molecule is added to the growing chain, the new end of the chain inherits the reactivity. Hence the molecule grows very rapidly.

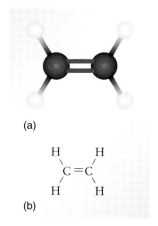

(a)

(b)

Figure 5.11
Ball-and-stick representation of ethene alongside the shorthand version of the same molecule.

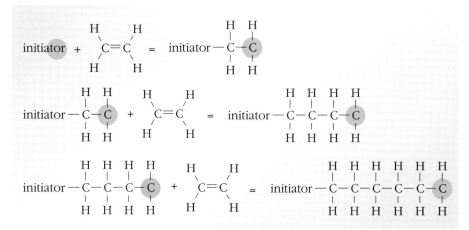

Figure 5.12
A representation of the polymerization of ethene. The highlighted part of each structure indicates the seat of reactivity. The chemical nature of the initiator is not important to our discussion.

This type of process is also called a 'chain reaction' and produces very long polymers. The chain building is stopped when two chains with reactive ends combine with each other or when an initiator combines with a reactive end of a chain, in either case producing polyethene with the formula initiator$(CH_2)_n$initiator, where n stands for a large number.

This method of forming a polymer chain is an example of **addition polymerization**, because the chain is built up by adding ethene monomers one at a time to the growing end. Notice the important point about polymerization reactions – to make a polymer, we must have a monomer with two ends that can take part in the reaction. Just as a railway carriage needs couplings at both ends to be part of a long train, so a monomer needs two ends that can be joined to other monomers. This principle also applies to a second method of making polymers, which we shall meet shortly.

Before we move on, there is another term to introduce. As we have seen, in polyethene the monomer is ethene. Once incorporated in the polymer chain, each ethene provides a small repeating unit known as a **monomer unit**.

▪ What, do you suggest, is the monomer unit in polyethene? (You should write the monomer unit with an 'unconnected bond' at each end.)

■ You know that the molecule is a long chain of $-CH_2-$ units, but CH_2 cannot be the monomer unit. Each ethene monomer provides *two* CH_2 units. Therefore the monomer unit is $-CH_2-CH_2-$, derived from the original ethene molecule in which one of the two carbon-to-carbon bonds has been split open to give a linking point at each end.

Whenever a monomer is linked into a polymer, the portion incorporated is termed the monomer unit. Having begun to develop ideas about the structure of polymers through polyethene, we can now consider some fibres.

5.3.2 We all need fibre

If you look closely at a fabric made from cotton or wool, you will see that it is made of very thin threads of varying length; these are fibres. Until relatively recently, all fibres used to make fabrics were obtained from natural sources – whether animal, vegetable or mineral. The principal sources may be animals (whence we get wool and silk) or plants (whence we get linen and cotton). Mineral fibres, such as asbestos, will not be considered here. All these fibres are called natural fibres. In recent years, we have been able to make (or synthesize) fibres; many of these are now familiar through their family names as in the case of polyesters.

Whether synthetic or natural, all fibres are similar in physical form. They are all fine threads many hundred times longer than they are thick. However, the different types of fibre have very different properties as we all know, and these differences are because the molecules that make up the fibres differ from each other chemically. These chemical differences are very important because they determine not only the characteristics of fibres and fabrics made from them, but also the uses to which fibres are put, and the way we treat them in washing and so on. We consider natural fibres in this Section.

The cotton plant is a tropical annual, which bears white or pink flowers. The fruit of the plant is called a 'boll' (a form of pod) which contains 20 or more seeds (Figure 5.13). Cotton fibres are the hairs that grow from the seed coat, and each seed may have as many as 20 000 of these fibres, which vary in length according to the type of plant grown in different countries. Cotton from the West Indies, for example, may have a fibre length of 6 cm, whereas North American cotton fibres are shorter (about 3.5 cm). Cotton from India

Figure 5.13
Cotton plant.

and China has even shorter fibres. When the bolls are ripe they are harvested for the cotton, which then goes through a process to separate the seeds, leaves and other unwanted material. The raw cotton is processed further by drawing out the fibres and spinning them into yarn.

Despite the physical differences in the size of cotton fibres from the various sources, they are all similar chemically. They all consist almost entirely of the polymer **cellulose**. The relatively small molecule glucose is the basic building block of cellulose, and in the next book of this Course you will learn more about this important substance. For now, all we need to know is that the glucose monomers react together to form a chain of cellulose. If you visualize the monomer unit from each glucose molecule as one of your imaginary paper-clips, cellulose resembles many hundreds of these linked together in a chain (Figure 5.14). Polymers like cellulose that are made by living organisms (that is, *bio*logically) are often called **biopolymers**.

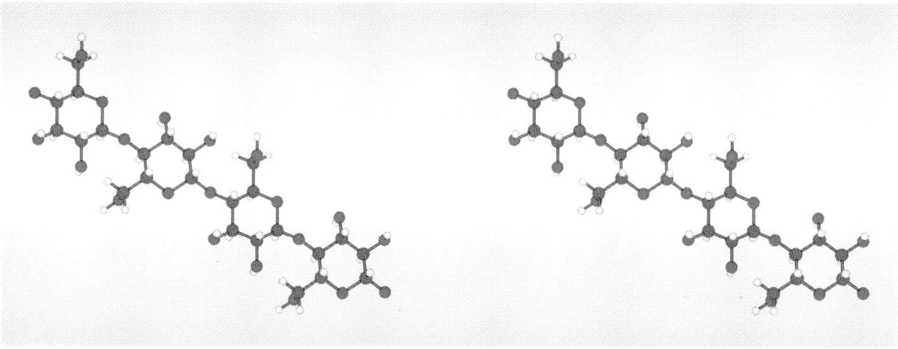

Figure 5.14
Stereostructures of a small section of cellulose.

Box 5.3 Exploding pinnies

One story illustrates how we can use chemistry to modify a natural polymer to give other useful properties. Christian Schönbein, Chemistry Professor at the University of Basle, accidentally broke a bottle containing a mixture of concentrated sulfuric acid and nitric acid. To prevent this potent mixture causing too much damage, Schönbein grabbed a cotton apron to wipe the acid from the floor. He rinsed the apron and then dried it by the oven – it quickly ignited. Thus, in 1846, Schönbein had by chance converted cotton into gun cotton, also called nitrocellulose or cellulose nitrate – which was ideal for the explosives industry.

Cellulose nitrate behaves very differently from cellulose. For example, it dissolves in organic solvents to give a sticky liquid, which can produce a protective layer over wounds. This material, collodium or collodion, was the first sticking plaster.

On gentle heating, cellulose nitrate can be moulded into hard objects. It can also produce thin films – the first moving pictures were recorded on cellulose nitrate films – hence we have the term 'films'. A mixture of cellulose nitrate and camphor provided an ideal replacement for ivory in the manufacture of billiard balls. The cellulose nitrate/ camphor mixture is known as celluloid.

Now you should prepare for some of your own experiments. They are the type of investigation that may be done by a chemical analyst or forensic scientist.

Experiments 2.10–2.13

First, carry out the test for cellulose on a sample of cotton, as described in the experimental notes (Experiment 2.10). Make a careful record of your observations.

Repeat the test for cellulose on a sample of linen fibre (Experiment 2.11). Once again, note down your observations.

Now carry out Experiment 2.12 to investigate the composition of wool.

Finally, carry out Experiment 2.13 to see if you can confirm the absence of sulfur in silk.

Linen, another natural fibre, comes from the flax plant. It consists primarily of cellulose, but it also contains up to 25% of other substances, such as waxes, which give the fibre a glossy appearance and make it brittle.

You should have gathered sufficient experimental evidence to be able to confirm the similarities of cotton and linen. They both contain cellulose, as your iodine tests should have confirmed. You may also know that both fibres are flammable and burn with a smell of burning paper to leave a white ash. That the fibres give a smell of burning paper may give you a clue that paper is also made up of the natural polymer, cellulose. You may have concluded from your first experiments that you cannot distinguish between cotton and linen by these chemical tests, and this should not be surprising because the polymers are chemically similar. Fortunately, many fibres can be differentiated by other techniques, such as using a microscope to examine their appearance (Figure 5.15).

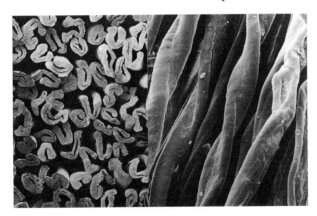 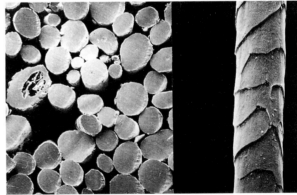

Figure 5.15
Close-ups of cotton fibres (left) and wool fibres (right). Notice that they are readily distinguished by their different appearances.

Wool is an animal fibre. The earliest clothes worn by our ancestors probably were the pelts of animals killed for food. The hair covering the skin of many animals comprises an overcoat of long smooth hairs and an undercoat of shorter, softer hairs. The hairs of the undercoat of sheep become matted or felted, and by removing them and causing them to felt artificially, the first non-woven wool fabrics were produced. The fleece of sheep has a thick undercoat, and the hairs of other animals, such as goats, camels and rabbits, also serve alone or as blends to produce fabrics with special properties. Wool from any source can be shown to consist almost entirely of protein molecules. In contrast to cellulose, which comprises chains of glucose monomer units, proteins are long chains derived from monomers called amino acids. These monomers and the biopolymers obtained from them are essential to living things and will be dealt with in more detail in Book 3.

You should have found from your experiments that wool contains sulfur. This is because some of the amino acid monomers contain sulfur and they take it with them as they become monomer units in the protein structure. As a result, wool contains sulfur atoms in adjacent protein molecules. In particular, the monomer unit from an amino acid called cysteine contains a sulfur atom. If a pair of these sulfur atoms are close together, they are able to form a sulfur-to-sulfur bond as a bridge between different wool molecules. These bonds are known as **cross-links** and help to maintain the shape of the fibres by locking together adjacent molecules, preventing the chains from being completely mobile. In fact, the polymer chains in wool are coiled in a spiral-like spring like the cable on a telephone. The sulfur bridges survive any stretching of the fibre and ensure that it reverts to its original shape once the tension has been released, thus contributing to the ability of wool fabrics to retain their form. We shall look at cross-links more closely later.

Interestingly and unfortunately, the same cysteine that causes the cross-links in wool also makes the fabric ideal as food for moth larvae. Silk is also a protein fibre, but does not contain cysteine, and is not usually attacked by moths. Silk is believed to have been woven into cloth by the Chinese for over 4 000 years. Like wool, silk is of animal origin, but comes from the cocoon of the silk moth.

In its life cycle, the silk moth passes through a larval stage. The larva is called a silkworm (Figure 5.16), and when it reaches about 8 cm long it is ready to pupate. A thread is spun from the spinneret (a mouth appendage), and the worm winds this thread around itself many hundred times. The silk thread is continuous and may be up to 2 km long. The worm uses chiefly two amino acid monomers to build the protein called 'fibroin' that makes up the raw silk filament.

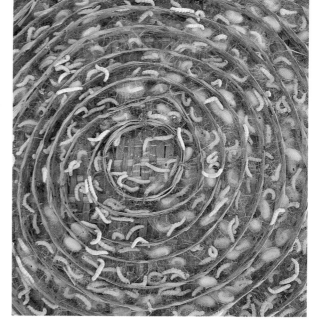

Figure 5.16
Silkworms.

Unlike the situation with wool, the chains of fibroin are not coiled, but are held together in a zig-zag arrangement by hydrogen bonds, giving rigid sheets. In this sheet arrangement, the chains are nearly fully extended. The structural differences explain why silk is less elastic than wool, less attractive to moths, and smoother to the touch. The sheets of protein are loosely stacked in silk and so can slide over each other with ease, giving a smooth and flexible fabric.

You may already know that both wool and silk will burn in a flame. The fumes given off have a smell that has been described as of 'burning feathers' and they do not have a sharp smell like vinegar. Other hair fibres behave similarly, and this behaviour therefore helps us distinguish silk and hair fibres from other fibres.

5.4 Tailor-made polymers

If, as humans, we are going to rival the endeavours of silkworms and cotton plants and provide ourselves with polymers for specific purposes, we are going to have to understand and control the underpinning chemistry. This Section aims to illustrate some aspects of tailor-making synthetic polymers. Again, polyethene provides a useful starting point. Let's look at the polymerization of ethene in more detail to assess opportunities for controlling the reaction to give a product with particularly desirable characteristics.

5.4.1 Polyethene and related plastics

There are two distinct types of polyethene: high-density polyethene (HDPE) and low-density polyethene (LDPE). At the macroscopic level, the differences between the two types are clear (Figure 5.17). The low density version is flexible yet tough; it is completely resistant to chemicals, but is sensitive to high temperatures. We commonly meet the high density version in the form of moulded bottles for applications such as bleach and large volumes of milk. The rigidity of HDPE allows even thin-walled bottles to keep their shape when filled with milk and stacked, and when exposed to even more demanding conditions (Figure 5.17). To appreciate the reason behind this difference in behaviour at the macroscopic level, we need once again to look in more detail at the chemical structure.

First, remember that we have to think about structures in three dimensions. High-density polyethene is a straightforward linear chain of monomer units like our linked paper-clips. It is called a straight-chain alkane. You should note that the terms straight-chain and linear refer to the fact that all of the carbon atoms in the constituent chains lie in a row, −C−C−C−C−. (In reality, chains that have a carbon 'backbone' actually have a zigzag shape, and applying the term straight-chain or linear should not be taken to deny that known shape.) In a branched-chain alkane, there are one or more carbon side-chains attached to the main chain. Figure 5.18 should jog your memory that you met some branched-chain alkanes in Book 1; the isomers of hexane did not have all six carbon atoms in a linear chain but instead they had branches.

Figure 5.17
The effect of heat on bottles made from HDPE (right) and LDPE (left).

Figure 5.18
The linear alkane, hexane, and one of its branched-chain isomers.

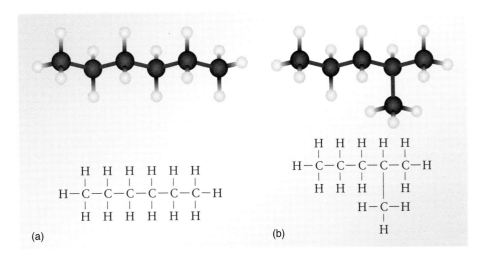

(a)

(b)

Because of the lack of branching in straight-chain polyethene molecules, they can pack closely together. There are definite regions in the polymer where the chains lie in an orderly form. These regions of regular and close packing are called **crystallites** and the polymer is said to have crystalline regions. In between the crystallites are regions in which the chains are more randomly ordered with respect to one another. The crystallites act like cross-links, holding different polymer chains together tenaciously, and account for the high density and great strength of HDPE. In contrast, low density polyethene is really quite a mess at the molecular level. Instead of a linear chain, it has branches sticking out of the main chain; that is, it is a branched-chain alkane. The branches prevent the chains packing closely together – hence it has an open structure with few crystallites, resulting in a low density product. You can imagine that polyethene producers would want their methods of making polyethene to supply the correct form for the job in hand, so control of the chemical conditions is vital to their business.

If ethene is polymerized at high pressure (1 000 bar) and relatively high temperature (200 °C), a waxy solid is formed in which the chains have irregular branches. The polymer product is therefore LDPE. This was the first form of polyethene to be manufactured. In 1953, Karl Ziegler observed that when certain metallic compounds were present in the polymerization process, they promoted the formation of long straight-chain polymers under milder conditions..Pressures of 1–10 bar and temperatures of 50–100 °C could be employed. As a result of these discoveries, the new and tough polymer HDPE became available. Substances like the metallic compounds used by Ziegler, which are able to make chemical reactions proceed more easily than in their absence, are known as catalysts. These will be covered in more detail in Book 3.

For his discovery, Ziegler shared the 1963 Nobel Prize for Chemistry with Giulio Natta, who showed that the catalyst system was widely applicable to other addition polymerizations of monomers containing double bonds. Now, we examine briefly some more addition polymerizations.

Ethene is not the only molecule that contains a carbon-to-carbon double bond and that undergoes addition polymerization. We can think of replacing one of its hydrogen atoms with another atom or group. By making this substitution and carrying out a polymerization, we have changed the chemical make-up of the product and thereby produced a different polymer with different properties. In this way, we can design polymers for specific uses. In the following examples, the main point to notice is the common feature of the double bond in the different monomers.

If we replace a hydrogen atom in the original ethene skeleton with a chlorine atom, we produce the substance called chloroethene, although it is more commonly known as vinyl chloride.

▪ Look back at Figure 5.11 and write the structure of the molecule formed by replacing a hydrogen atom in ethene with a chlorine atom.

▪ Your representation could be a ball-and-stick structure or a simplified structure as in Figure 5.19.

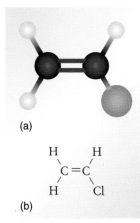

(a)

(b)

Figure 5.19
Ball-and-stick
representation of
chloroethene (vinyl chloride)
alongside the shorthand
version of the same
molecule.

The most important feature of the vinyl chloride molecule, $CH_2{=}CHCl$, is that it still has the double bond in its structure, and so can add to other molecules of the same type to produce poly(vinyl chloride) or PVC (Box 5.4). You will recall that ethene is a flat molecule with fixed geometry about its double bond; vinyl chloride is similar.

▨   What will be the monomer unit in PVC?

■   The monomer unit in PVC is $-CH_2-CHCl-$ derived from the original monomer in which one of the two carbon-to-carbon bonds has been split open to give a linking point at each end.

When writing the structure of the PVC monomer unit as $-CH_2-CHCl-$, you should not imagine that the bond between monomer units involves the chlorine atom. It is carbon atoms that form the backbone of all such polymers.

Notice an important point about this type of structure: the two carbon atoms of the monomer unit have different atoms attached to them. It is said to be asymmetrical. Let us call the CH_2 group the head of the monomer unit and CHCl its tail.

▨   In principle, asymmetrical monomer units can add to each other tail-to-head or head-to-head. Draw, or use your model kit to make, portions of PVC that show the different structural possibilities.

■   We could have chlorine atoms on alternate carbon atoms or on adjacent ones as in Figure 5.20.

In practice, addition polymerization of chloroethene and similar substituted ethenes occurs in head-to-tail fashion, and the resulting polymer chain has the substituent atom or group on alternate carbon atoms as in Figure 5.20a. This is not the end of the subtleties regarding spatial arrangements of atoms in polymers like PVC, but such detail is not covered in this Course.

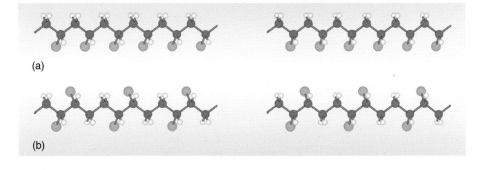

(a)

(b)

Figure 5.20
Stereostructures of
poly(vinyl chloride) chains
formed by (a) tail-to-head
addition and (b) head-to-
head addition.

Box 5.4 PVC: production and problems

Over the decade to 1991, worldwide production of PVC increased from 10 million to almost 15 million tonnes a year. This made it second in production tonnage to polyethene (23 million tonnes), with the main applications being in a range of products for the building industry, such as window frames, piping and floor coverings.

It is remarkable that the monomer for PVC, chloroethene, is very toxic, whereas the polymer is essentially non-toxic. Stability to environmental degradation is an obvious requirement for use in building products, but this same property presents difficulties when it comes to disposal. The principal disposal route for waste materials in the UK is to bury them in landfill sites, and the lack of biodegradability is a disadvantage here. Incineration is another option, but the presence of chlorine in PVC means that air pollutants are produced when the plastic is burned.

To widen our discussion of different examples of addition polymerizations, it is useful to know a little about the molecular structure of a substance called benzene. This is discussed in Box 5.5.

Box 5.5 Benzene and the phenyl group

Carbon atoms can form rings as well as chains. Benzene is an important compound that provides an example of a type of carbon ring in which the bonding is somewhat complicated. It comprises six carbon atoms joined together in a planar ring with one hydrogen atom attached to each of the six carbon atoms. Its molecular formula is C_6H_6, and a simple way of representing the

arrangement is as a hexagon of carbon atoms with alternate single and double bonds joining them, and a hydrogen atom joined to each carbon (Figure 5.21). Notice that this structure gives each carbon its four bonds and each hydrogen one bond. At this stage of the Course, there is no need to probe the structure of benzene any further.

The **phenyl group** is simply benzene with one

Figure 5.21
Representing benzene.

of its hydrogen atoms missing. It is not a molecule; it cannot be stored or bought. Rather, it is a part of an organic

structure, just as Cl— is part of vinyl chloride. The formula of the phenyl group is C_6H_5-, where the unconnected bond shows that it is available for attachment to another group to form a molecule. One example is styrene, best known as the monomer for polystyrene. Such compounds based on the structure of benzene are termed **aromatic**.

If instead of chlorine we substitute a phenyl group (see Box 5.5) into the ethene molecule, we produce styrene (Figure 5.22). This is the basic building block of polystyrene, the well-known material of drinking cups and packaging. In styrene it is the $CH_2=CHC_6H_5$ double bond that is active in the addition polymerization; the special nature of the double bonds inside the benzene ring structure prevents them taking part in the reaction.

■ Now write down the monomer unit in polystyrene.

■ You should have written a structure like $-CH_2-CHC_6H_5-$ or you may have drawn out the phenyl group as a ring.

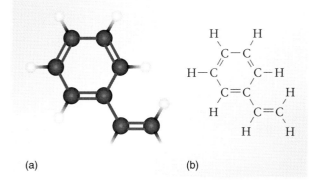

Figure 5.22
(a) Ball-and-stick model of styrene, the monomer of polystyrene; (b) a shorthand representation of the same molecule.

Methyl methacrylate is a complicated monomer yet it undergoes exactly the same addition polymerization as does ethene, vinyl chloride and styrene, and forms poly(methyl methacrylate). You are probably more familiar with this polymer as perspex, which is used for making baths and as a substitute for glass in hard contact lenses (Box 5.6).

Box 5.6 Eye contact

One example of the medical applications of polymers is in the production of contact lenses. Glass was used for many years, with each contact lens being ground individually, but about 50 years ago poly(methyl methacrylate), or perspex, lenses were introduced. Perspex is tough, easily machined to make contact lenses, and possesses good optical properties. Thus, it made an attractive replacement for glass. Most important, however, is its physiological inactivity in the eye. It has been said that this inactivity was discovered by chance when it was found during the Second World War that fighter pilots were able to endure splinters of perspex in their eyes when their cockpits had been damaged. Unfortunately, though, hard contact lenses often caused physical irritation of the surface of the eye, and new materials for contact lenses have therefore been designed to avoid this effect.

An example of this chemical design is provided by the polymer poly(2-hydroxyethyl methacrylate). As its name suggests, this material is related to poly(methyl methacrylate).

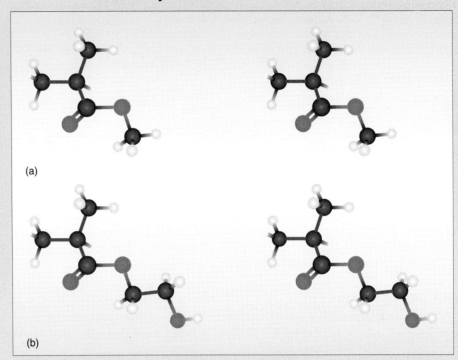

Figure 5.23
Two relatives — stereoscopic representations of the monomer units in (a) poly(methyl methacrylate) and (b) poly(2-hydroxyethyl methacrylate).

■ Look at Figure 5.23 and summarize the difference between the monomer units of these two polymers.

■ Poly(methyl methacrylate) has a CH_3 group capping the larger branch off the main chain, whereas in poly (2-hydroxyethyl methacrylate) the CH_3 group has been replaced by CH_2—CH_2—OH.

Note in particular that poly(2-hydroxyethyl methacrylate) contains an —OH group, which we call the hydroxyl group. Both of the polymers are normally hard, clear and glassy, but when water is present, the presence of the hydroxyl group in the newer material becomes significant and brings into play the important role of hydrogen bonding. As water also contains hydroxyl groups, this common feature allows the polymer to take up water, which it retains by hydrogen bonding. This converts the polymer into a clear flexible gel, rather like a film of water covering the eye. The absorbed water effectively acts as a **plasticizer** - a molecule that lubricates the movement of polymer chains against each other. The overall result is that the lens is less irritating. Such polymers used in making soft contact lenses are called **hydrogels**. The flexibility of hydrogels also is convenient in restoring sight after the removal of cataracts; the lenses can be folded and inserted through a smaller incision than would be required for a hard lens.

Several accidental discoveries occurred during the polymer revolution. One took place in 1938 when a research scientist by the name of Plunkett needed about 50 kg of a gas, tetrafluoroethene (that is, ethene with the four hydrogen atoms replaced by four fluorine atoms, $CF_2=CF_2$). He synthesized this substance and stored it in steel cylinders. However, he found that the tetrafluoroethene did not flow from the cylinders. On cutting them open, no gas escaped but he discovered a white powder instead. This was polytetrafluoroethene (PTFE or Teflon).

▨   What is the monomer unit in Teflon?

▪   You should identify that it is $-CF_2-CF_2-$. This is just like the monomer unit in polyethene, but with H replaced by F.

Teflon was the outcome of an accidental discovery in terms of being unplanned, but was noticed by a receptive mind. It was found that the surface of Teflon was heat resistant, very difficult to wet (just as water does not wet a highly waxed car body) and presented low friction. Teflon is therefore the ideal coating to produce non-stick cookware.

5.4.2 Synthetic fibres

We are now in a position to consider synthetic fibres. Until the end of the 19th century, the only fibres used to make textiles were the natural fibres wool, silk, cotton and linen, which we described previously. However, as early as 1664, the naturalist Robert Hooke had predicted that one day we might be able to imitate the action of the silkworm and produce silk artificially. To Hooke, it must have appeared that the silkworm merely made a thick sticky liquid from its food (mulberry leaves), and forced this liquid through its spinneret so that when exposed to air it hardened to give a silk thread. In fact, it took over 200 years for us to mimic the silkworm.

The first factory to produce artificial silk commercially was opened in 1889 by Count Hilaire de Chardonnet. He had been working with Pasteur to investigate diseases of silkworms and so became interested in ways of producing silk. Cellulose makes up the cell walls of plants, and Chardonnet used cellulose taken from mulberry leaves. Remember also that Schönbein had found that cellulose could react with nitric acid to give cellulose nitrate, and when this dissolved in organic solvents, it formed a sticky liquid that could be drawn into threads.

However, Chardonnet was the first to use a 'spinneret' – a metal plate with small holes in it through which the solution of cellulose nitrate could be forced (Figure 5.24). Many filaments could be produced simultaneously and continuously, and this made artificial silk a commercial possibility. However, high flammability was a disadvantage of the so-called Chardonnet silk – you will recall that cellulose nitrate was used as an explosive! Later, to reduce this hazard, the cellulose nitrate filaments were 'denitrated' to leave regenerated filaments of pure cellulose. As a fibre, Chardonnet silk resembled natural silk but, as you should appreciate by now, it has a very different fundamental structure. Chemically, Chardonnet was barking up the wrong tree – or plant! His Chardonnet silk was essentially the plant polymer cellulose, whereas silk is a protein polymer.

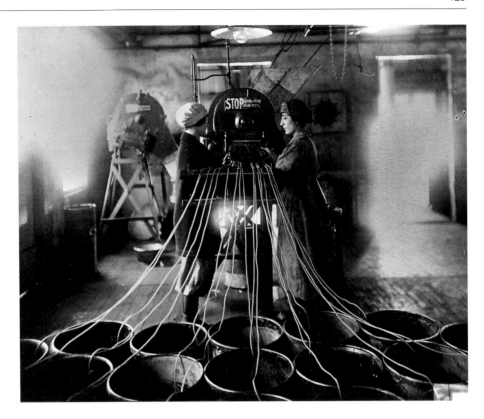

Figure 5.24
Cellulose nitrate from spinnerets.

Nowadays, his process is no longer used, but he nevertheless laid the foundations of the vast synthetic fibre industry. The synthetic fibres now produced from cellulose are known collectively by names such as 'viscose' and 'rayon'.

In order to produce synthetic polymer fibres, we need to have a regular molecular structure, some degree of order in the way the chains are arranged, and strong intermolecular forces, like hydrogen bonds, between the chains. We will now see how these properties are obtained in a variety of products, taking polyesters as our first example.

In AV *'Organic chemistry 1'*, you learned that certain groups of atoms within a given molecule tend to be active in reactions and that the rest of the structure remains inert throughout the reaction. The reactive groups are called **functional groups**. Here, we do not need to develop the idea of functional groups to any degree but we need to reinforce the principle that molecules contain inert parts and reactive parts to understand another method of producing polymers, called **condensation polymerization**.

▪ Cast your mind back to earlier comments about the most important property that a monomer must have in order to build up a polymer chain. Then suggest how many reactive functional groups should occur in a monomer.

▪ Earlier, we likened a monomer to a railway coach. To take part in a long train, the coach must have couplings at each end. Similarly, the monomer must have *two* functional groups, one at each end of the molecule.

As long as an organic molecule has two suitable functional groups, then it will participate in the polymerization. It so happens that there are several types of reaction that enable monomers to join together and, in the process, release water as a by-product. The formation of water droplets on a cold surface is termed condensation, so this polymerization reaction acquired the title of condensation polymerization.

This process should be contrasted with addition polymerization, in which the monomers simply add to each other and do not produce a by-product of any sort. Also in addition polymerization, all monomers need to contain only a carbon-to-carbon double bond. We did not have to consider any further functional groups. With condensation polymerization, there is a wide range of different functional groups that can participate.

All functional groups, including the connecting groups that join monomers together via a condensation polymerization, have names. Table 5.2 has already revealed the names of four important connecting groups: esters, amides, carbonates and urethanes. These are the names of groups formed when monomers react together in a condensation reaction to give a polymer. In other words, depending on the type of polymer, one of these four connecting groups becomes an integral part of the polymer chain. If polymers were trains, the couplings that hold the coaches together would be of four types: ester, amide, carbonate or urethane.

Figure 5.25 shows the structure of the commonly used polymer, poly(ethylene terephthalate) (PET). The chemical structure looks complicated, but the important thing is to focus on the chain of monomer units and functional groups that hold it together. This particular example is called a polyester because the chain contains connecting groups called **esters**. The ester groups are highlighted in the diagram. Any polymer containing this type of link in its main chain would fall in the category of polyester.

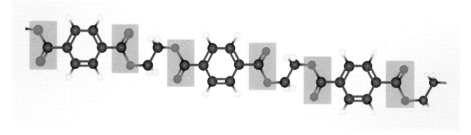

Figure 5.25
The structure of a typical polyester, poly(ethylene terephthalate) (PET). It is used in making boil-in-the-bag food packaging, cassette tape and extensively for soft-drink bottles. In fibrous form, this polyester is commonly known as Terylene, Crimplene or Dacron. In the structure, the ester groups are highlighted in grey.

Nylon is a very well known synthetic polymer. It was invented by Wallace Carothers (Box 5.7) and his colleagues, and first announced to the public on 27 October, 1938.

Box 5.7 Wallace Carothers

Born in 1896, Carothers (Figure 5.26) graduated in accountancy and secretarial studies, and subsequently studied and then taught chemistry at various universities, including Harvard. In 1928, he left the academic world to lead a research group in the chemical company Du Pont, which unusually for that time was embarking on a programme of basic research. While at Harvard, Carothers began thinking about polymers and eventually defined goals of preparing molecules of known structure by using known chemical reactions, and investigating how properties of the molecules depended on their make up. Within two years of joining Du Pont, he had described his work in many patents and scientific papers, and in 1935 he prepared a polymer fibre with the best balance of properties and manufacturing costs. Despite wide recognition as an outstanding scientist, Carothers was obsessed with the thought that he was a failure, and committed suicide, unfortunately before large-scale manufacture of nylon had begun.

Figure 5.26
Wallace Carothers (1896–1937), the inventor of nylon.

His discovery was timely, because the new synthetic fibre entered the marketplace just in time to replace silk, which was in short supply during the Second World War. Not only did nylon replace silk in stockings, but also parachutes and tent fabric for the war effort were made from it. Nowadays, nylon has wider applications than in fibres alone and is one of the foremost examples of the impact of chemistry on our daily lives.

Nylon is made when the appropriate monomers with two functional groups are used to form a long chain via a condensation polymerization. A portion of its structure is shown in Figure 5.27. The actual polymer chain may be made up of over 20 000 monomer units. The connecting group in nylon, highlighted in Figure 5.27, is called an **amide** group. Unlike the ester group, it contains a nitrogen atom.

▪   Which family name would you give to nylon?

▪   The monomers are joined through amide links, so nylon is a polyamide.

Figure 5.27
The structure of nylon. It is a strong synthetic fibre and resists abrasion. On washing it does not shrink or stretch, but it is degraded by ultraviolet light so it is not used for curtains. In the structure, the amide groups are highlighted in grey.

Many different polyamides may be manufactured, with the objective of producing products to suit particular needs. Different polyamides are made using different starting materials, but all of the monomers share the property of forming two amide links. The amide link is also found in hair, and this explains why burning nylon smells like burning hair. In fact, all proteins are polyamides. Book 3 includes more detail on these biopolymers, which are essential chemicals in living things.

Question 35 The structures of acrylonitrile, propane and propene (propylene) are given below. Which will form polymers? Would condensation or addition polymerization be appropriate? Write out the monomer units for the polymers formed.

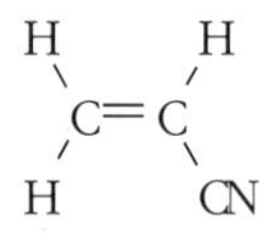

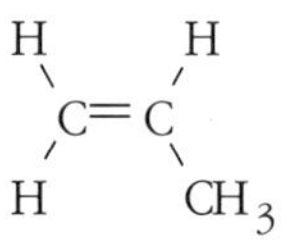

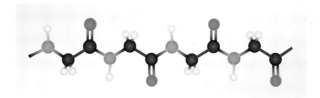

Question 36 Summarize the main differences between condensation polymerization and addition polymerization. Examine the structure of the polymer given in Figure 5.28. Which method of polymerization was used to prepare it?

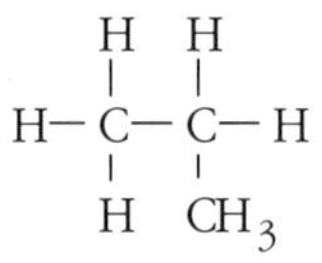

Figure 5.28
The structure of a polymer, for use with Question 36.

5.5 Properties of polymers

So far, we have concentrated on the naming, structure and production of polymers. Some of their properties have emerged in passing. We now take a more coherent look at properties of polymers.

Before continuing with this text, carry out Home Experiment 2.14.

Experiment 2.14

As you do the experiments, think about some of the peculiar properties of polymers that they demonstrate, and consider how these properties relate to polymer structures. As you try the experiments, observe carefully and write notes. The explanations for the effects should become clearer as you continue.

Box 5.8 Some terms in polymer technology

Extrusion – In a meat mincer, a screw forces meat through small holes. A polymer extruder works similarly. Plastic granules are heated and then forced through a metal nozzle or 'die'.

Blow moulding – As in making a bottle. A divided bottle-shaped mould is clamped around an extruded plastic pipe with one end sealed. Blowing compressed air into the pipe expands it like blowing up a balloon, and it takes the shape of the mould.

Injection moulding – This is like using a giant syringe. Molten plastic is forced through a hole into the mould, from which the article is removed after cooling.

These techniques are illustrated diagrammatically in Figure 5.29.

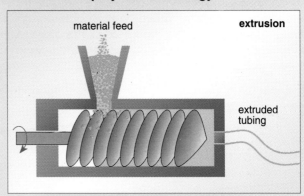

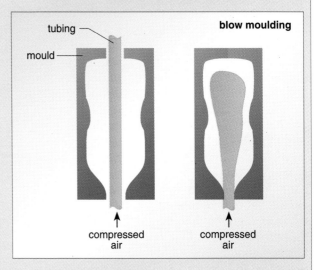

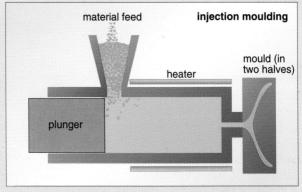

Figure 5.29
Processing techniques in polymer technology.

5.5.1 Why is plastic wrapping film hard to tear?

In the Home Experiment that you did with the strip of polyethene, you should have found that at first the entire strip elongated very slightly and then part of it became narrower. Technically, we say it formed a 'neck'. Then you should have noticed that the neck became longer and material apparently flowed into it from the taut regions on either side, so these became shorter. Eventually, the neck grows to the whole length of the strip, which becomes noticeably stiffer before it breaks. We can represent this observation graphically. In engineering terms, the force we exerted per unit original (or undeformed) area of polymer is called the stress, and the change in length can be described as the strain. A graph showing the relationship of stress and strain in the polyethene strip as it is stretched is shown in Figure 5.30.

You may have noticed that the effort you needed at the start was greater than that needed to continue the stretching process. This is shown by the rise in the graph followed by the decline after the initial peak. Let's examine what was happening at the molecular level during this deformation process (Figure 5.31).

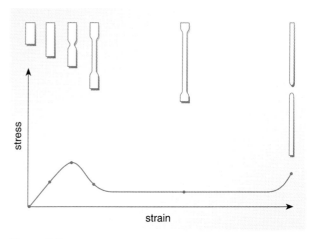

Figure 5.30
A typical stress–strain curve for stretching a strip of polyethene.

To the left of the Figure, the molecules are coiled randomly, not under stress. As we apply and increase the stress, the molecules start to line up in the direction of the applied load. They can do this because the stress is sufficient to overcome the weak forces between the molecular chains. The force we applied is not strong enough to break the covalent bonds between the carbon atoms, but becomes sufficiently strong to draw the molecules into an aligned state. This is what happens in producing fibres, as we shall see. Eventually, the molecules become highly aligned and the stress is borne entirely by the crystallites that hold together the aligned polymer chains. Finally, we apply sufficient stress to disrupt even the crystalline regions and the individual molecules slide apart. As a result, the strip breaks. The difficulty in making the break in your sample of polyethene may have reminded you of the infuriating difficulty that we have all experienced when opening a plastic bag of frozen peas or similar product. We can understand why this is the case when we remember that the long polymer molecules are coiled partly at random

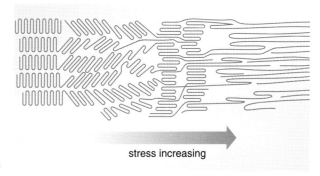

Figure 5.31
Deformation at the molecular level.

but have crystalline regions where the chains are strongly held together. We can easily displace a coiled mass because the forces holding the coils together are not strong. Suppose we now arrange these polymer molecules so that most of them lie in the same direction. We can do this by extruding (or forcing) molten polymer through holes and allowing the material to harden, when the molecules will usually align themselves along the long axis. In this aligned form, there is an extensive crystalline region so the polymer molecules are not easily torn apart.

Many manufacturers of plastic wrapping now introduce a convenient flaw into the packaging to concentrate the stress and so make the bag easier to open. You should now begin to appreciate how these everyday activities relate to the molecular structure of polymers.

5.5.2 Cross-linking

The synthetic polymers or 'plastics' we have met so far can all be reshaped by high temperatures and/or pressure; they are examples of thermoplastics. In general, they are softer and less brittle than thermosetting polymers. High temperatures enable the molecules in thermoplastics to move relative to each other, especially when a mechanical stress is applied as when we extrude a fibre. This occurs because the weak forces between the polymer chains are easily broken. If we make the forces between the linear chains stronger, then we can expect the behaviour to be different.

Up to this point, we have mostly envisaged a polymer as a string of monomer units joined end to end. This arrangement is a straight-chain polymer. Suppose that during polymerization we introduce a small quantity of a monomer that not only has reactive groups at each end, but also has one in the middle.

▨   What effect might this have?

▩   Introducing the new monomer with three reactive ends will allow branching. Instead of linear molecules, chains can form on the sides of the main chain, and we can also link chains together through these branches (Figure 5.32).

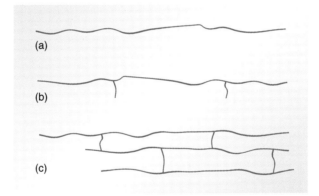

Figure 5.32
Highly schematic view of branching and cross-linking. The Figure shows (a) a straight-chain polymer as a string, (b) a branched-chain polymer, and (c) a cross-linked polymer in which the branches bond together adjacent molecules.

It is said that cross-links are formed, a term introduced previously. You met it last in the context of proteins like wool and hair. The amino acid monomer cysteine has two functional groups, which allow it to be incorporated into a protein molecule, but it has one reactive end 'left over'.

▨   Summarize what you recall about cross-linking through cysteine monomer units in proteins.

▩   You learned that the sulfur atom in a cysteine monomer unit of one chain can combine with the sulfur atom of another cysteine monomer unit in an adjacent chain. The resulting sulfur-to-sulfur bond helps to lock the two protein chains together.

Box 5.9 Tanning and perming

Cross-linking is an ancient art known in several different forms before its molecular basis was understood, and each art used its own terminology.

Cross-linking in skins to form leather was 'tanning', cross-linking of paints containing linseed oil was 'drying', and cross-linking of resins was 'hardening'.

The natural sulfur bridges that cross-link hair can be broken by chemical treatment. The hair can then be stretched and held while sulfur bridges reform to lock the hair into a more fashionable shape. The process is called 'permanent waving' or just a 'perm'.

In summary, cross-linking offers the possibility of three-dimensional arrangements of polymer chains held together by strong covalent bonds.

Question 37 What effect do you suppose that strong covalent bonds forming cross-links in a polymer will have on its physical properties?

One of the earliest synthetic cross-linked polymers was bakelite. It is prepared by condensation polymerization between chemicals called formaldehyde and phenol. First, a more pliable product is made when there is little cross-linking and this is moulded into the required shape. Because the phenol molecule has three reactive sites, branching and cross-linking can occur. Accordingly, when the moulded material is heated, further chemical cross-links are formed resulting in a hard, durable product. Melamine and Formica are similar thermosetting polymers.

5.5.3 Why is rubber rubbery?

We are all familiar with perspex as a clear, glass-like polymer or as the material from which many baths are made, but heat it above 105 °C and it becomes pliant like rubber. Conversely, rubber becomes hard and brittle when made very cold. So what makes rubber rubbery?

Natural rubber is a unique polymer material with a long history. In the late 16th century, a Spanish historian noted:

> *Christopher Columbus, having again landed on the island of Hispaniola, on his second voyage to the New World in 1493, set out to explore his new domain, search for gold and observe the natives, whom he called Indians, since he still believed he had reached the [East] Indies. In their villages he noted that one of their pastimes was a game played with balls made of the gum of a tree, which tho' heavy would fly and bound better than those fill'd with Wind in Spain.*

(Journal of Chemical Education, 1990)

Rubber was brought to Europe from the Amazon region in 1735 under the name of 'caoutchouc', derived from the Indian words 'caa' (wood) and 'o-chu' (to flow or weep). The famous chemist Joseph Priestley (who amongst his other activities discovered oxygen) gave the material the name India rubber in 1770, after he observed that this material from the West Indies erased pencil marks on paper rather better than using breadcrumbs as previously.

Priestley apparently ignored the discovery by Charles de la Condamine that the Amazon people used the sap from the rubber trees for waterproofing garments and for making waterproof boots by pouring the sap on their feet. One of the most important applications of rubber came in 1823 when Charles Macintosh made a solution of rubber which enabled a sandwich of rubber between two layers of cloth to be made. Unfortunately, natural rubber had the disadvantage of hardening and cracking when cold, and softening when warm.

In 1826, another famous chemist, Michael Faraday (as seen on £20 notes!) found that rubber was built up from basic building blocks called isoprene with a chemical formula C_5H_8. Being a chain of isoprene monomer units, rubber could be called polyisoprene. The monomer unit in natural rubber can be written in shorthand as:

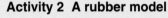

$$\begin{array}{c} CH_3 \\ | \\ -CH_2-C=CH-CH_2- \end{array}$$

Activity 2 A rubber model

Now use your model kit to make a three-dimensional model of the monomer unit in rubber. Note in particular if you can arrange the model of the monomer unit in different ways. In other words, is there only one possible arrangement of the atoms in the monomer unit or are there more than one?

   You should find that you can arrange your model in two distinct ways. Examine the model with the double bond horizontal and the groups around it pointing up and down. In one arrangement, the unconnected bonds that represent links to the rest of the chain are on the same side of the double bond – either both up or both down. This is called a *cis* arrangement (pronounced 'sis', from the Latin for 'on the same side'), and we call the product *cis*-polyisoprene (Figure 5.33). Natural

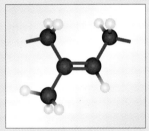

Figure 5.33
Ball-and-stick representation of the monomer unit in *cis*-polyisoprene.

rubber has many thousands of isoprene units joined in the *cis* arrangement. Notice particularly what we mean by 'on the same side'. The chain links are not both on the same carbon atom of the double bond. They are attached to each carbon atom of the double bond and, when that double bond is drawn horizontally, both chain links lie either above or below the level of the double bond.
The other possibility is for

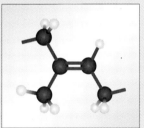

Figure 5.34
Ball-and-stick representation of the monomer unit in *trans*-polyisoprene.

the links to be on opposite sides of the double bond as represented in Figure 5.34. This is called *trans*-polyisoprene, from '*trans*' meaning 'opposite'.

■ Is it possible to convert the *cis* monomer unit into the *trans* monomer unit simply by rotation about the double bond? (Hint: you may need to think back to Book 1 and the discussion of the properties of double bonds.)

■ No. You learned in Book 1 that rotation about a single bond occurs readily but rotation about a double bond does not normally happen. Thus, the *cis* and *trans* models cannot be readily interconverted and the resulting polymers, *cis*- and *trans*-polyisoprene are quite distinct.

■ What is the name given to molecules that have the same composition (in terms of the numbers of different atoms) but exhibit different spatial arrangements of their atoms and hence different properties?

■ They are called isomers (Book 1, Section 7.5).

In the case of polyisoprene, the *cis*-isomer occurs as natural rubber; *trans*-polyisoprene also occurs naturally but as the hard, horn-like polymers gutta percha and balata. Gutta percha comes from the leaves of *Palaquium* trees from Malaysia and the East Indies, and balata comes from *Mimusops globosa*, grown in northern South America, and was used as golf ball covers. Gutta percha was the first material used for solid golf balls; until its introduction, golfers used the 'feathery' ball (leather stuffed with feathers).

In natural rubber, we have the basic requirements for 'rubberiness', which we can define as the ability to be deformed considerably and yet recover completely. We also use the term **elastomer**, particularly for synthetic rubbers. Rubberiness requires first and foremost that the material must consist of polymer chains. In unstretched rubber, the tangled polymer chains are held together weakly in coils by London forces. Stretching the rubber makes the polymer molecules change from a coil to extended chains. Once the stress is relaxed, the polymer molecules spring back to their original shape as long as they have not slipped over each other. Natural rubber does not regain its original shape completely because significant molecular slippage does occur. The problem is solved by cross-linking the chains in a process called **vulcanization** (Box 5.10). Joining perhaps 1% of the segments of a polymer chain allows the chains to be elongated and aligned under stress, but prevents the extended chains from sliding irreversibly over each other. This markedly improves the elasticity of rubber.

▦ Why does rubber become stiff and rigid when it contains many cross-links?

■ When there are lots of cross-links in rubber, there are many short connections linking the main polymer chains. They are much less free to move relative to each other.

▦ What will be the result if the cross-links between the chains are more widely spaced?

■ There will be fewer connections linking the main polymer chains and so more movement of the chains between the cross-links will be allowed.

▦ Which product is more extensively cross-linked: vehicle tyres or elastic bands?

■ Rubber tyres are more rigid than elastic bands because in tyres there is more cross-linking.

Nowadays, much rubber is synthetic. Names like butyl rubber (widely used for linings for such things as tubeless tyres, ponds, and at landfill sites to prevent toxic liquids reaching underground aquifers), polybutadiene (used for golf ball covers) and SBR rubber (vehicle tyres) may be familiar to you. Although these materials are structural variations of natural rubber, their polymer chains also exist as a very flexible tangle.

You may also be familiar with silicone rubber as door and window sealants, artificial heart valves and other medical applications. We do not

Box 5.10 Vulcanization

Chains of natural rubber are very long and have few if any cross-links, and so the material is a thermoplastic, becoming soft and sticky in the summer and hard and brittle in the winter. These disadvantages were overcome in 1839 by a discovery made accidentally by Charles Goodyear in Woburn, Massachusetts. The story goes that after many years of experimenting, he had spilt or accidentally placed a mixture of rubber, sulfur and lead oxide on a hot stove. The rubber was no longer sticky but had been converted to a tough, elastic substance stable to heat and cold. It also did not dissolve in the solvents that dissolved natural rubber. He had invented the process now known as vulcanization (Figure 5.35).

Vulcanization is a chemical reaction between sulfur and rubber resulting in cross-links being formed between the rubber polymer chains. Notice from Figure 5.33

Figure 5.35
Goodyear's discovery.

that there are double bonds present in the polymer molecule. You should remember that double bonds provide a major route to the formation of polymers, so it should not be a surprise to find that these double bonds can serve to provide covalent links between the chains. We can form covalent bonds between the polymer molecules, and if we do this the material will become much more rigid because the chains are no longer free to move apart. The more cross-links between chains, the more rigid the rubber

until eventually the polymer is so cross-linked that it is no longer rubbery because there is no flexibility of the chains between the cross-links. Goodyear's vulcanization process produces a controlled amount of cross-linking. The sulfur reacts with the double bonds and forms sulfur bridges as cross-links between the chains, resulting in a huge three-dimensional network (Figure 5.36). The covalent cross-links survive the stretching and help the molecules to spring back once the tension has been relaxed. This type of network molecular structure lies behind the explanation of why rubber is rubbery.

■ Which type of natural polymer has sulfur cross-linking?

■ Previously, we mentioned sulfur cross-links in proteins like wool. Then we explained how this made wool elastic.

Many synthetic and natural fibres can be

stretched, as we have seen in the making of polymer fibres (Figure 5.31). The main difference between rubber and other fibres is that rubber goes back to its previous shape and size. We saw from Figure 5.31 that stretching a fibre aligns the polymer chains. In fibres, this alignment allows forces such as those of hydrogen bonding between chains to have an increased effect, and they will be strong enough to hold the fibres in their stretched, aligned position. In rubbery polymers (elastomers), we find there are large and bulky groups along the chains, and these prevent the chains from packing together so closely. As the chains are further apart, there are not the same forces between them to keep them in the uncoiled arrangement. This weak interaction between polymer molecules is not enough to keep the rubber in its stretched position so it reverts to the original coiled state.

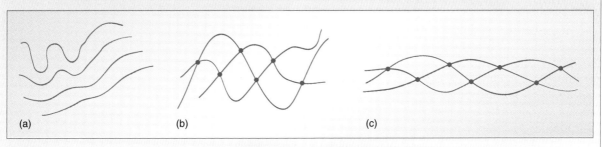

Figure 5.36
Cross-links in rubber. (a) Unvulcanized natural rubber molecules have few if any cross-links; (b) vulcanized rubber has a network structure with cross-links; (c) vulcanized rubber on stretching.

have the space here to cover this material in detail, but you should note that chemically it represents a break with tradition. Instead of being based on a string of carbon atoms as in the case of conventional addition polymers, the 'backbone' consists of alternating silicon and oxygen atoms.

5.5.4 Changes of state

From Chapter 6 in Book 1, you should recall the three states of matter: solid, liquid and gas. Water provides an example that everyone can readily appreciate.

If you observed the properties of a pure sample of water over a range of temperatures, you would see marked changes at the melting and boiling temperatures, that is, whenever it undergoes a transition between two states. Ice melts rapidly to a mobile liquid when the temperature exceeds its melting temperature and, above its boiling temperature, a sample of liquid water gives a larger volume of steam.

If you were to repeat this experiment with a material such as glass, you would find a different behaviour. Instead of a sudden change corresponding to melting, we would find simply a gradual change from a hard solid to a treacle-like substance. The temperature at which this starts to occur is known as the **glass transition temperature**. Above the glass transition temperature, the material is more like treacle, whereas below it the material appears as a solid. The change is therefore more subtle than melting.

So, water freezes rapidly when the temperature falls below its melting temperature, but certain other materials cannot assume ordered, crystalline arrangements as readily. For example, honey – a concentrated aqueous solution of the sugars saccharose, glucose and fructose – takes a long time to crystallize because the larger molecules take time to arrange themselves. For polymers, there is greater difficulty. When a molten polymer is cooled, the chains cannot easily find the best arrangement, and the problem is worse if there are irregularities in the chains. So polymers tend to produce non-crystalline or only partially crystalline arrangements. Nevertheless, even in the amorphous (or unarranged) region, the polymer chains are coupled, and one chain cannot be moved without adjacent chains moving. This is rather like trying to move one strand of cooked spaghetti in a bowlful without moving the other strands. This co-operative movement of polymer chains is very restricted below the glass transition temperature. Remember that it is not 'melting', but a transformation of a glassy body into a viscous material.

As an example, we can consider chewing gum, which is often based on poly(vinyl acetate). With a glass transition temperature of around 28 °C, it is solid when purchased but softens in the mouth (which will be about 37 °C). Glass transition temperatures vary widely for different polymers because they depend partly on the length and stiffness of the polymer chains and partly on the interactions between the chains. As the latter property is also influenced by the history of the polymer (whether it has previously been drawn, softened, extruded, etc.), there can also be differences in glass transition temperatures quoted in reference tables, even for the same polymer.

Cotton consists of cellulose molecules with a glass transition temperature of 225 °C. As the cellulose chains cannot move significantly at room temperature, cotton fabrics keep their shape. Cotton also absorbs water easily (because they both contain −OH groups – yet another effect of hydrogen bonding) and so is ideal for underwear as it absorbs perspiration. The water acts as a plasticizer or lubricant, allowing the chains to move, and so the glass transition temperature is lowered to 20 °C. Then, as the chains can move, the fabrics tend to crease when uneven pressure is applied. Ironing removes the creases, but a damp cloth or steam iron makes the movement of the fibres easier.

5.6 Selected applications

We have already mentioned the role of polymers in the production of contact lenses, but there are many other applications in medicine. Dentistry is one such area where polymers find wide use. Mature teeth are biologically less complex than many tissues, and so polymers can be used for restoration or for bonding other materials to teeth.

For filling teeth, silver amalgam (an alloy containing tin and mercury) is commonly used for back teeth as it has been for many years, but cosmetic preferences dictate that front teeth be filled with tooth-coloured materials. It has been found that addition polymers can be used for fillings, but a common feature is the use of visible blue light for initiating the polymerization process once the filling is in place in the tooth.

Denture bases comprise the largest group of laboratory-made dental polymers. These are usually made by moulding a dough of poly(methyl methacrylate). In the future, we are likely to see more polymers designed to meet the special requirements of dentistry.

There are other demanding applications for custom-built polymers. The chemistry of polymers has been driven to a large extent by the need for materials to satisfy specific applications. Strong materials must stay strong, flexible materials must remain flexible, and so on. Unfortunately, polymers are nothing more than organic molecules and so, for example, the bonds holding the molecules together can be broken, thus reducing the chain length, or cross-links may be broken. One way of enhancing the stability of commercial products against the ravages of weather, light or ageing is to add stabilizers. For example, compounds of tin, lead, cadmium or zinc are among the stabilizers used in PVC products. Plasticizers are used in products in which internal movement of polymer chains is needed, as described previously in relation to contact lenses and cotton.

Several ideas have been suggested for explaining the effects of plasticizers. Perhaps the simplest has already been mentioned; that is, that the plasticizer acts as a lubricant and allows the polymer chains to move more freely. Another possibility is that the plasticizer molecules disrupt the interactions between polymer chains and so free up the movement of those chains.

A variety of complex molecules may serve as plasticizers, but because there are no covalent bonds between them and the polymer chains, they can migrate to the surface and be lost by abrasion, solution or evaporation – so the product eventually becomes brittle.

PVC on its own is rigid and difficult to process, but it is widely used in piping and construction materials, including window frames. Rigidity is an obvious requirement for these applications so unplasticized PVC or UPVC (also known as PVC-U) is used. Small amounts of plasticizer allow PVC to be rolled into sheets for floor coverings. Much higher contents of plasticizer (perhaps 50% by mass) make a soft and flexible product for use in wire insulation, shower curtains, outdoor clothing and upholstery. The use of plasticizers in the 'vinyl' for car interiors used to become apparent after they evaporated on hot days and produced a film of plasticizer condensate inside the car.

The demanding applications of the aerospace industry have driven the development of a number of polymers able to withstand very challenging conditions. High strength-to-mass ratio and high temperature resistance are but two of the properties essential for such applications. It will probably be no surprise that polymers, most of which are organic, usually soften relatively easily and many burn. The problem was to find out which chemical bonds were likely to contribute to these weaknesses; the challenge was then to replace the weaker links with stronger ones.

Perhaps the largest group of thermally stable polymers is a group of polyamides, which are manufactured readily and have been successful in many practical applications. You will recall that polymers such as nylon are known generically as polyamides, and are manufactured from monomers using a condensation reaction. When benzene rings are incorporated into the chains, the polymers are known as 'aramids', after 'aromatic', which is the term applied to the family of compounds that are based on the structure of benzene. Such materials are made by condensation polymerizations in a similar manner to nylon, and commercial products include Nomex and Kevlar (Figure 5.37). Note that the structure of Kevlar is composed essentially of rigid benzene rings held together by amide groups – the same type of connecting group that we saw in nylon.

Figure 5.37
A molecular model of a portion of Kevlar.

Let's think why Kevlar molecules adopt an ordered structure. You should notice that the molecule contains an —NH—C=O group, which can form hydrogen bonds. Therefore, dipole–dipole forces and hydrogen bonds hold the molecules securely aligned, like logs in a river. The resulting Kevlar fibres are stiff and strong. Another factor operating involves bonding along the chain as a result of the presence of benzene rings, and this also makes each fibre more like a rod.

Applications of these materials include products as diverse as body armour (bullet-proof garments), skis, golf clubs, aircraft structures, yachting sails and as replacements for asbestos in gaskets and brake linings.

On a mass-for-mass basis, Kevlar is five times stronger than steel because it is built up from lighter elements. Its great strength is due to the way that the rigid straight-chain molecules align themselves parallel to each other, giving an ordered structure in the form of sheets of molecules.

Question 38 Mark the squares in Table 5.5 to indicate which terms on the left refer to the substances shown at the top.

Table 5.5 For use with Question 38.

Term	Terylene	Polyethene	PVC	Polyisoprene	Nylon	Bakelite
condensation polymer	✓	✗			✓	✓
cross-linked polymer						✓
thermosetting plastic						✓
polyamide					✓	
thermoplastic polymer	✓	✓	✓	✓	✓	
addition polymer		✓	✓	✓		
polyester	✓					

Let's move on to the polymer that you are currently examining. You are very familiar with paper, but have you ever thought of it as a polymer, still less as an aggregate of fibres? Examine a piece of paper closely and you will see its grainy surface. Tear a piece of paper – preferably not from this Book – and the fibrous structure will be revealed at the edges.

The cell walls of all plants are made of cellulose.

▪ What is the monomer used by nature to make cellulose?

▪ You may remember that cellulose is a polymer made from glucose molecules.

We have seen that plants, the source of several natural fibres, consist of a major proportion of cellulose, and wood is no exception. The typical composition of wood is about 45% cellulose, 25% lignin (another polymer) and 20% of hemi-cellulose (a branched-chain polymer), with the balance being resins, etc., which are characteristic of the type of tree.

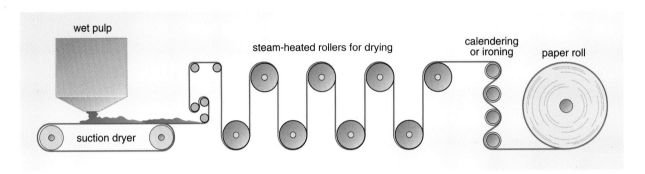

Figure 5.38
Manufacture of paper.

In a simple paper-making process (Figure 5.38), logs are torn apart mechanically in water, which carries away much of the lignin and non-polymer material to leave pulp – a mixture of cellulose in water. Better quality papers are made by chemical pulping, which involves cooking wood chips with chemical solutions. These break down the lignin to leave cellulose and hemi-cellulose fibres. Chemical treatment of the pulp then produces various grades of paper.

Figure 5.14 shows the chemical structure of the glucose monomer units in cellulose. It is obviously a complex structure and the details need not be remembered, but note that the monomer units are joined together through oxygen atoms.

Bacteria in herbivorous animals can break down this oxygen link and so thrive on cellulose, but we humans cannot do so. Another point you should notice about the chemical structure of cellulose is that there are OH groups attached to the ring structures. If you remember from Book 1 that the uneven electron arrangement in OH groups leads to hydrogen bonding, you should understand why cellulose molecules can form hydrogen bonds to each other. There can be several hydrogen bonds to each glucose monomer unit and, as a result of this hydrogen bonding between the chains, cellulose is stiff and does not stretch. However, the presence of OH groups means that pure cellulose paper will absorb water readily and form hydrogen bonds to it. This can break down the structure and this is what happens to tissues that are almost pure cellulose. Other papers have additives (perhaps 10% of their mass) to produce more robust products with specific properties. A chemical problem is to blend the additives and cellulose to match the needs of the application.

There is much interest nowadays in recycling paper, whether motivated by the desire to preserve forests or to reduce the amount of waste paper for disposal. Specialized additives in certain paper products make them unsuitable for recycling. Printing ink may be difficult to remove and may accumulate as more and more paper is recycled, so making paper increasingly grey in colour. More significant is the fact that the fibres break down on recycling, and the decrease in length of the fibres means that the strength of recycled paper decreases. Despite these problems, much paper is recycled into a variety of products, as a visit to your local supermarket will show.

5.7 Reuse, recycle or break down?

A major advantage of polymers is their stability, but eventually products become outdated, stop working or wear out, and the question of disposal has to be considered. What do we do with the potentially vast amounts of waste? The versatile properties and low cost of polymers have revolutionized our lives to the extent that worldwide over 100 million tonnes of synthetic polymers are produced each year. One of the advantages of synthetic polymers is their durability – they don't break down in the environment in the same way as natural fibres rot, for example. This durability presents society with problems. Of course we can bury them in a landfill site where they will remain almost indefinitely, or we can send them to an incinerator and take advantage of their combustibility, but are there other options? Throughout history, we have reused waste and scrap materials whenever practicable, so why should polymers be an exception?

In fact, we can often reuse or recycle the polymers or their building blocks. Another option is to design the molecules so that they do break down in the environment to give harmless products. However, the choice between which of the options is best in the long term is not an easy one.

Plastics are often viewed as materials presenting environmental problems in terms of the amount of waste they constitute (Figure 5.39), their disposal problems, difficulties in recycling, and the amounts of natural resources required for their production. Let's examine some of these views to see if they are supported by facts.

We are familiar with the amount of packaging on all goods that we buy nowadays. About one-third of all packaging in the UK is made of plastics, and a third of all plastics goes in to packaging with over half being used for food and drink (Table 5.6).

1% non-ferrous metals
4% textiles
7% plastics
9% ferrous metals
11% putrescibles
12% glass
21% miscellaneous
35% paper

15 million tonnes of household waste arise in the UK each year

Figure 5.39
Waste mountains.

Table 5.6 Common plastics used for packaging.

Polymer	Packaging applications
low density polythene	sacks, bags, dustbin liners, squeezy bottles
high density polythene	bottles for bleach, milk and fruit juice; bottle caps
polypropylene	margarine tubs, crisp packets, packaging film, sauce bottles, paint containers
PVC	blister packs, food trays, shampoo and squash bottles
polystyrene	egg cartons, yoghurt pots, vending machine cups, bottles
PET	carbonated drinks bottles, oven-ready food containers and trays

We have commented that plastics are derived largely from crude oil, and about 4% of oil production goes to plastics manufacture, whether in raw material or energy terms.

Polymers seem to be a major part of the waste packaging that we throw away. Certainly, plastics occupy a large volume. In the 1990s, typically 30% of household waste, but they constitute only about 7% of domestic waste by mass according to both UK and US statistics, whereas paper and cardboard make up about 40% by mass. Textiles constitute about 8% of US household waste. Overall, then, polymers do represent a major part of the rubbish we throw away, but let's focus particularly on synthetic polymers and compare plastic packaging with paper packaging. Studies from Germany indicate that producing supermarket carrier bags made of paper costs more than twice as much in monetary terms and in energy consumption than the production of plastic bags. Similarly, making polystyrene containers uses less energy than similar bleached cardboard products. Polythene bags also contribute between 74% and 80% less by volume of solid waste for disposal than paper bags. Many other statistics have been quoted. For example, to make one million bottles of one-litre capacity from PVC we need about 66 tonnes of oil, whereas to make the same number of glass bottles we use about 230 tonnes of oil. Similarly, 1 million square metres of polypropylene film needs about 76 tonnes of oil compared with the 178 tonnes needed to make the same area of cellophane from wood. In the long term, we have to find ways of preserving natural resources and of reducing pollution, such as carbon dioxide from burning energy resources.

Much of the plastic waste can be recycled in theory, but until recently little plastic recycling was done. For example, in 1989 plastic recycling from municipal waste in Europe amounted to 4%, and in the US the figure was a mere 1%. However, legislation on the recycling of 'post-consumer' packaging is on the increase, and this trend is likely to continue. Fundamental to this trend is the separation of different types of plastic. We have seen the wide variety that is available, and separation of wastes according to polymer type is essential to maintain the highest recycle value. At the beginning of this Chapter on polymers, we asked you to identify polymers in your home. You may have found this difficult, but fortunately many polymers are now being labelled to aid identification and to assist in their separation for recycling. The Society of Plastics Industry has developed a system of labelling plastic items to facilitate recycling. If you examine a number of plastic containers, bottles or plastic bags, you are likely to see the appropriate symbols shown in Figure 5.40. Marking of products with symbols of polymer type is a step towards making possible more recycling of polymers in the future.

Figure 5.40
Codes on plastic packaging for ease of recycling. See Tables 5.1 and 5.2 for explanations of letters.

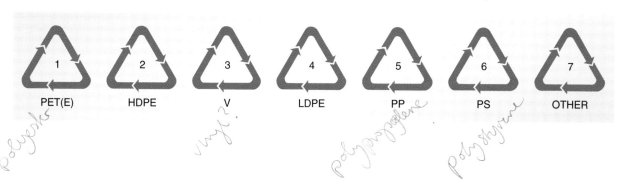

What do we mean by recycling polymers? Several options are possible. We can recycle or 'reuse' the product in a similar manner to the reuse of glass milk bottles, or we can recycle the material from which the article is made – perhaps by melting the article and making a new product identical with the original. Many materials are already being recycled. Glass recycling is commonplace, and in fact the inclusion of broken glass (cullet) in a glass-melting furnace is beneficial to the production process. Paper is recycled in relatively large proportion, as we have considered previously. A significant percentage of aluminium cans are also recycled, and this reflects the driving force of economics – the cost savings over making aluminium from its ores. So how much polymeric material could we recycle?

We have seen that much plastic is used as packaging, with about 35% being film products which are difficult to separate and recycle. Coatings and closures are even more difficult to collect, but bottles account for about half of packaging and can readily be collected for recycling – in theory at least! The lightness of polymer bottles can be a disadvantage if we need to collect 20 000 in order to make one tonne of 'new plastic'. The overall balance in resources for recovery and storage may outweigh the resource consumption in making virgin polymer from crude oil products. This may argue in favour of recycling the energy content of polymer waste by burning the material and recovering the heat to warm buildings or to generate steam for operating machinery. However, burning plastics is not without its problems.

As the use of synthetic fibres and plastics has increased, so has the incidence of accidents associated with fires involving these materials. Polymers burn by a process rather like that in a candle flame. Heat from the flame is radiated back to the polymer surface to release gases, which burn above the surface by a series of chemical reactions – you will learn more about this process in the next part of this Book dealing with energy.

Many injuries and deaths occur each year as a result of such fires, but it is not so much the fire but rather the smoke and toxic gases that represent the major hazard. The majority of fire victims die from inhalation of the combustion products of burning polymers. The most common toxic gases found in the combustion products are carbon monoxide, hydrogen cyanide, oxides of nitrogen, hydrogen chloride and hydrogen sulfide, as well as a variety of organic compounds. You're sure to have noticed the fire safety labels on furniture, for example, and the developments in this area have come from the serious risk of fires of upholstery foams. Similarly, fabrics have been made more resistant to burning.

Approaches to introducing fire retardance in fabrics and other polymers include:

- using fibres with inherent fire resistance;
- applying fire-retardant finishes to fabrics;
- adding inorganic fillers such as clays, alumina and silica;
- adding organic fire retardants containing elements such as chlorine, phosphorus and boron.

In general, many fire retardants work by raising the temperature at which decomposition of the polymer begins, whereas others interfere with the series of chemical reactions taking place in burning and so inhibit the process. Other strategies rely on forming a protective layer of char on the surface of the polymer, and this layer effectively insulates the surface from the heat radiating from the flame.

Notice from our list above that the presence of chlorine in a molecule suggests that it becomes more difficult to burn. So it is with PVC, which burns only with difficulty and extinguishes itself when taken out of a flame. Unfortunately, when chlorinated polymers do burn, the combustion products may be particularly hazardous. There have been numerous problems with 'cable-burning' – recovering copper from old electrical wiring by burning off PVC insulation on an open bonfire.

The toxic products produced when many polymers burn raises not only concerns about safety within the home, but also concerns about environmental harm when polymers are incinerated at the end of their life. The combustion gases from waste incinerators must be treated to prevent the toxic gases from burning plastics escaping to the atmosphere.

There are many other opportunities for dealing with polymer waste rather than merely burning or burying it in the ground. Wearing lemonade bottles may not appeal, but one idea to keep millions of plastic drinks bottles from landfill involves converting them into clothes – giving a whole new meaning to that bottle-green sweater. The process is essentially simple. Bottles are chopped up, impurities removed and the polyester extruded into a fibre, which is spun into yarn and then knitted with virgin polyester to produce fabric.

Figure 5.2 at the beginning of this Chapter showed the variety of polymer components in a typical car. Plastic materials constitute about 10% of the total mass of a modern car, and this proportion is likely to increase because their lower densities relative to those of metals allow vehicles to be lighter and so more fuel-efficient. Plastics also offer better resistance to corrosion, good thermal and noise insulation, and enable components with complicated shapes or many functions to be fabricated more easily than from metals. In recent years, many car manufacturers have looked increasingly at opportunities for using recycled materials. Some applications are relatively straightforward, such as the use of macerated fabrics in sound-proofing materials for floor and roof-linings. When it comes to recycling the polymer components from cars themselves, a balance has to be drawn between economic and environmental benefits as well as quality and reliability. Thus, there may be difficulties in recycling the material of a given component into another of the same type. Consequently, new bumpers from old may not be viable, but bumpers can become wheel-arch linings or heater ducting, and wheel-arch linings may in turn become boot linings and floor insulation material. Thus, we have a recycling option of remelting and using the material for less demanding applications than the original.

A new approach to recycling polymers took a step forward in 1991 with the approval by the US Food and Drug Administration of recycled plastic containers to be used in contact with food. The breakthrough in technology supporting this move involved chemically breaking down PET polymer into its original monomers. PET is the trade name for poly(ethylene terephthalate) which is a polyester (Table 5.2) and it is the polymer used in many plastic drinks bottles. It is produced by condensation polymerization, which has the by-product of water. In the new recycling strategy, the notion is to reverse the polymerization reaction and so it can be thought of as treating PET with water to regenerate the monomers. The feedstock for the breaking down process can be PET unsorted by colour and containing contaminants typically present in post-consumer waste. Once the original monomers have been regenerated they are used to make fresh resin, from which new containers are extruded.

This example of chemistry at work for recycling essentially creates a new supply of raw material from what we used to 'dump' in a landfill site as waste. By closing the loop, it is an excellent example of one of the fundamental laws of chemistry – the **law of conservation of matter**. Thus, while atoms may change partners, they are not destroyed by chemical reactions, and the same numbers of atoms remain at the end of a reaction as were present at the beginning.

Rather than using industrial chemistry to break down polymers, there has always appeared to be an attraction in developing certain polymers that will break down harmlessly in the environment in a manner similar to that of natural fibres and paper. There are two impediments to the widespread use of biodegradable polymers. While we may want packaging and similar products to degrade, we do not want it to occur too rapidly nor do we want degradable polymers to be recycled with non-degradable polymers; just imagine the catastrophic implications if degradable polymers were recycled into your new PVC window frames!

However, biodegradable polymers do have a number of important potential uses (Box 5.11), including the production of medical implants for pharmaceutical applications, as in slow-release capsules for medicines.

> **Question 39** What problem do you envisage in attempting to recycle a thermosetting polymer along the routes suggested so far (including Box 5.11, overleaf)?

With so many possible routes to recycling, it is inevitable that we ask ourselves which is the best. Is it better to recycle polymers, to make them degradable or simply to discard them after use and make new products from virgin resources? This question has vexed many people for a long time and becomes more critical as concerns over sustainable development grow and the costs of waste disposal escalate. One tool that may help resolve the problem is described in the following case study, which ends this Chapter.

Box 5.11 Biodegradable polymers

One biodegradable polymer, Biopol, was introduced by ICI in the late 1980s, although it has links to a polymer that has long been known – polyhydroxybutyrate (PHB) – produced from a monomer called hydroxybutanoic acid in a condensation polymerization. A polymer related to Biopol is used for soluble sutures (stitches).

The polymer PHB was first discovered in 1926 and is a brittle material. Another product can be formed from a member of the same series of monomers, but with one more carbon atom. This compound is 3-hydroxypentanoic acid. Once again, this has two reactive functional groups and can form a polyester. By making a polymer that incorporates both of these monomers, a more flexible and thermoplastic product can be formed.

An unusual feature about this polymer made from a mixture of two monomers (called a co-polymer) is that it is produced commercially by a fermentation process. A naturally occurring bacterium *Alcaligenes eutrophus* is introduced into a medium containing glucose and other substances, including pentanoic acid. Fermentation breaks down the feedstock to produce the desired product, which accumulates in the bacteria as they grow. To put it simply, the polymer accumulates in the bacteria as a food reserve in a manner similar to the way we accumulate fats, and at the end of the process the polymer may make up 80% of the dry weight of the bacteria. Harvesting the cells and treating them allows the polymer to be released for purification. The product known commercially as Biopol has been processed to produce bottles and fibres. For example, one of its first uses was to package a range of shampoos.

■ What implications can you see for polymers like Biopol in relation to sustainable development for the future?

■ A particular advantage of such a polymer is that it is produced from renewable resources like sugars, rather than from limited reserves of crude oil. Sugars can also be produced locally rather than depending on imported raw materials. The biodegradability reduces disposal problems.

If you remember that the polymer is derived from compounds present in living systems, the biodegradability of Biopol should not be too surprising. The first step probably involves the breaking of the ester bonds between the monomer units by the action of water, and the smaller molecules so released are then converted into carbon dioxide and water by normal biological oxidation processes. Other biodegradable polymers incorporate starch, another natural product. So, a polymer of polythene containing some starch can be broken down into smaller residues by bacteria attacking the starch component.

Case study – getting to the bottom of an environmental engineering problem!

One thing that all of us almost certainly have in common is that we have worn nappies. No matter whether the traditional cotton reusable type or the more modern disposable version, polymers are the construction materials of these essential garments. Nappies are used in a rather hostile environment. They must withstand much movement (stretching and pulling) and are exposed to aqueous solutions containing acids, bases and a mixture of organic compounds, some of which are solid. At the same time, nappies must be soft and safe – no toxic or irritant substances would be tolerated in such sensitive situations.

We saw previously that cotton consists almost entirely of cellulose, built up of a long chain of about 3 000 or more glucose monomer units, and so a cotton nappy is relatively simple in construction! A disposable nappy is more complex, so let's see how it's designed to satisfy the demands we make of it. First and foremost, a nappy must be absorbent: it must retain a great deal of liquid and for this property it is common to use the same polymer as in the traditional nappy – namely cellulose. However, cellulose for disposable nappies is obtained from wood pulp rather than from cotton. Look back at Figure 5.14 and notice the many hydroxyl groups (−OH) in cellulose. This commonality to water should immediately suggest to you the possibility of hydrogen bonding. The cellulose fibres are cut quite short and are packed together to give a large absorbent surface area. The fibres are held in place between a layer of woven cellulose fibres that allows water to pass through, and an outer waterproof layer that stops leaking (hopefully!). This outermost layer must also be resistant to the forces of tension and compression that it will inevitably suffer, and a sheet of polythene serves admirably. It is waterproof, non-toxic and colourless, as well as being resistant to mild acids and bases, and also quite inexpensive.

One area of potential leakage is where a nappy fits around the waist and legs. A close yet flexible fit may be obtained by using elasticated sides to a nappy. Polymers providing this elasticity are typically related to rubber, which you may remember is *cis*-polyisoprene. Disposable nappies also replace the traditional nappy pin by adhesive tabs. A polypropylene sheet is somewhat stiffer than polythene and serves as the backing to the tabs, while the adhesive is also a polymer – perhaps a polyisobutene of low to intermediate relative molecular mass which is sticky but not as strong as superglue, because we need to take off the nappy. This widely used product therefore illustrates how the combination of different properties of polymers, both natural and synthetic, can satisfy a commercial need.

From an environmental impact perspective, you may wonder whether disposable or reusable nappies are preferable. To try to answer this difficult question, a technique known as **life cycle analysis** (LCA) has been used. Definitions of LCA differ in detail, but the Society of Environmental Toxicology and Chemistry (SETAC) definition is:

> *Life cycle analysis is an objective process to evaluate the environmental burdens associated with a product, process or activity by identifying and quantifying energy and materials used and wastes released to the environment, and to evaluate and implement opportunities to effect environmental improvements. The assessment includes the entire life cycle of the product, process or activity, encompassing extraction and processing of raw materials, manufacturing, transportation and distribution, use/re-use/maintenance, recycling and final disposal.*
>
> *(Fava, J.A. et al. 'A technical framework for life cycle assessments', Report of the Workshop organized by the Society of Environmental Toxicology and Chemistry, Aug. 18–23 (1990), Smugglers Notch, USA.)*

Much initial work with LCA has focused on packaging materials, for which the amount of waste and issues such as recycling have assumed great importance. The first LCA was claimed to have been done in 1969 for Coca-Cola beverage containers, while a second in 1970 compared polystyrene foam and moulded pulp meat trays – both polymers, you will notice. Compared with many products, packaging is relatively simple and so convenient for quantitative environmental analysis, but this is not to suggest that LCA is easy. In terms of amassing the data, the interpretation of the results and in using those results, LCA is not without its difficulties. For example, the Advertising Standards Authority has criticized certain claims that disposable nappies cause no more damage to the environment than reusable ones. The ruling impinges on the use of LCAs in advertising. Various LCAs comparing the two products have differed on issues such as the number of reusable nappies used, and on quantifying pollutants associated with the products. Some analyses have also failed to take account of the manufacture of pulp from wood, so ignoring the energy and chemicals involved in these processes and in forestry, while packaging and disposal issues have not always been dealt with fairly. The Authority's ruling noted that environmental preferability depended on the subjective interpretation of the relative importance to be attached to different criteria. Consequently, we cannot recommend which nappies are best at present. The technique of LCA is being used more and more, but as with many scientifically based methods, it is important that the results are used and interpreted carefully. Try to interpret the results given in the following question.

Question 40 As an example of the life cycle or holistic approach, we can compare the use of paper and polystyrene as the material for disposable hot drink cups. Intuitively we may expect the paper cup to be superior from an environmental point of view, but is it? Some clues to providing an answer to this question are given in Table 5.7. The table contains a lot of data but you should use them just to form a broad impression. From that, decide which type of cup you would favour.

Table 5.7 A comparison of hot drink cups. Adapted from T.E. Graedel, B.R. Allenby and P.B. Linhard (1993) 'Implementing industrial ecology', *IEEE Technology & Society Magazine*, Spring, pp. 18–26.

	Paper cup	Polystyrene cup
raw materials per cup:		
wood and bark (g)	33	0
petroleum (g)	4.1	3.2
finished mass (g)	10	11.5
relative price	2.5	1.0
per tonne of material:		
steam (kg)	900–1 200	5 000
power (kWh)[a]	980	120–180
cooling water (m³)	50	154
water effluent: per tonne:		
volume (m³)	50–190	0.5–2
suspended solids (kg)	35–60	trace
biochemical oxygen demand[b] (kg)	30–50	0.07
organic chlorine compounds (kg)	5–7	0
metal compounds (kg)	1–20	20

	Paper cup	Polystyrene cup
releases to air per tonne:		
chlorine (kg)	0.5	0
sulfides (kg)	2.0	0
particulate matter (kg)	5–15	0.1
pentane (kg)	0	35–50
recycling potential:		
primary user	possible	easy
after use	low	high
ultimate disposal:		
heat recovery (MJ/kg)^a	20	40
mass to landfill (g)	10.1	1.5
biodegradable	yes	no

^{a} Units of energy are considered in Part 2 of this Book.

^{b} Biochemical oxygen demand (BOD) is a measure of the pollution potential of an effluent passing to a river or of the pollution in a river. It is a measure of the amount of oxygen needed for microorganisms in the water to break down the pollution. The higher the BOD, the greater the pollution potential because more oxygen will be consumed from the water. As oxygen has only limited solubility in water, it is possible for the pollution to result in depletion of the dissolved oxygen, so making the water no longer able to support aquatic life.

Question 41 Before looking at the Summary, try this question. Polymers have many useful properties, but a variety of materials may be added to them in order to modify their features and to make them more suitable for industrial and consumer applications. Use information gained from this Chapter and your general knowledge to list some of these additives and their roles.

Summary of Chapter 5

Polymers are all around us in both the natural and manufactured environment. They all are very long molecules built up by joining together many smaller building blocks (monomers), which are often but not exclusively identical. Many polymers have traditional names based on their inventor's name or for other more curious reasons. Systematic names are more meaningful, and may indicate the building blocks from which the polymer is constructed, while other names indicate the type of connecting group between the building blocks. Polymers may be thermosetting or thermoplastic materials.

Natural fibres are natural polymers and have long been used by humans for producing materials such as fabrics and paper. Fibres of vegetable origin are based on cellulose, a polymer of glucose, whereas fibres of animal origin have a protein structure.

Synthetic fibres are manufactured by two principal routes – addition polymerization and condensation polymerization. Addition polymerizations involve the addition of monomer molecules to each other. The capacity for forming links arises from the breaking of one of the carbon-to-carbon bonds in the double bond that each monomer must have. The simplest addition polymer is polyethene, derived from ethene. Many other addition polymers are possible, and all may be regarded as deriving from substituting other groups into the ethene parent molecule. A second route for production involves condensation polymerization, in which a polymer is produced together with water. Many familiar synthetic fibres are produced by condensation polymerization, including polyamides and polyesters. A crucial property of a monomer is that it must have two reactive sites – rather like a railway carriage must have couplings at each end.

The three-dimensional molecular structure of polymers is important in relation to the desired properties of materials. The long chain structure influences the strength of polymers, but particularly important are the forces between the chains. These forces may be relatively weak – for example, London forces or hydrogen bonds – but nevertheless can still impart major influences on the behaviour of the materials. Stronger links between different polymer molecules are obtained through covalent bonds known as cross-links. These are important in relation to thermosetting polymers and in the behaviour of polymers known as elastomers. Polymers do not exhibit a melting temperature, but show a glass transition temperature.

The inherent stability of synthetic polymers is an advantage for many purposes over the ease of degradation offered by natural polymers, but this stability presents problems in disposal. Degradation in landfill sites does not take place unless the polymer is suitably designed, while destruction by incineration can release toxic products. With increasing concern over sustainable development and the conservation of resources, routes for recycling polymers are developing. While chemical methods to break down polymers to their raw materials are being introduced, recycling the materials themselves is being assisted by the greater use of identification labels, and the design of products for recyclability. Synthetic polymers can offer many environmental advantages over natural products, but the overall environmental impact must be assessed over the life cycle of the product.

Objectives for Part 1 of Book 2

When you have completed Book 2 Part 1, you should be able to:

1   Define and use the terms emboldened in the text in a correct context.

2   Relate the use of materials through history to the availability of raw materials and to developments in science and technology.

3   Outline the origins of materials in the Earth's crust and explain mechanisms for their concentration.

4   Indicate how extraction methods and costs of materials are related to the raw material location and the chemistry involved.

5   Describe the structure and composition of selected major materials.

6   Relate the chemistry of the production of cement to its use.

7   Relate the properties of glass to its composition.

8   Link the microstructure of ceramics to their physical properties.

9   Outline methods that can be used to control the reflective properties and the colour of glazes.

10   Rationalize the extraction methods for metals (specifically copper, iron and aluminium) with the position of metals in the activity series.

11   Explain the different corrosion tendencies of metals, and outline methods of minimizing corrosion.

12   Relate the properties of metal alloys to composition and atomic structure.

13   Represent, manipulate and extract data from a variety of sources including tables, graphs, pie charts and histograms.

14   Calculate the composition of substances in terms of proportion by mass and by volume.

15   Interpret chemical formulas of all types including empirical, molecular and unit cell formulas.

16   Given the structure of a unit cell, derive the cell formula and the empirical formula.

17   Outline the characteristics of the major structural types (ionic, molecular, giant extended, etc.).

18   Calculate concentrations from solubility and volume data.

19   Balance chemical equations.

20   Use chemical equations to calculate quantities of reactants and products.

21   Identify reactions involving acids or bases (Arrhenius definition) and oxidation and reduction.

22   Identify substances being oxidized and reduced in chemical reactions.

23   Appreciate and be able to identify polymers in the environment.

24   Relate examples of the historical developments in polymer chemistry, both by chance and by design.

25   Explain how changes in molecular structure of polymers have an impact on performance of materials in use.

26   Appreciate how different types of polymer are distinguished by their chemical structure and manufacture.

27   Recognize common polymer types, such as addition and condensation polymers, as well as the specific polymers featured in Chapter 5.

28   Discuss and give examples illustrating the significance of chain length, intermolecular forces, branching and cross-linking in polymers.

29   Explain the significance of ordered arrangements of molecules in the context of polymers.

30   Identify good fibre-forming characteristics in a polymer.

31   Outline in general terms the principles for making common polymer types.

32   Describe the role of additives in polymers and give examples.

33   Propose options for dealing with the increasing amounts of plastic waste generated by our modern lifestyles.

34   Appreciate the difficulties in balancing the environmental impact of natural versus synthetic polymers over their whole life cycle of use.

Comments on Activity

Activity 1

My list of polymers around me as I write these words is as follows.

My wooden desk, wooden window frame and pencil contain the polymer cellulose, which is also found in paper.

Veneer is held to the substrate of my desk by an adhesive, and the window frame and pencil are painted; paints are polymers.

My eraser is made of rubber, a natural polymer.

My carpet, sweater and shirt are all made of polymers: wool, cotton and a synthetic fibre (polyester), respectively.

Another window frame is PVC, which is clearly a plastic or polymer, as is my computer case.

My lunch-time sandwiches contain polymers – the starch in the bread and the protein in the beef – but protein polymers also form my skin and hair.

Your own list will be different, no doubt. It may be shorter or longer, depending how much you scoured your home for bin-liners, plastic buckets, cobwebs, electrical fittings, linen and so on. You may also have attempted to group polymers together as those that are strong, like work-surfaces, or flimsy, like food packaging. You may have noted a plethora of other characteristics, such as rigidity, elasticity, fibres, films and insulators. Such is the range of properties that can be 'programmed' into modern polymers. By the end of Chapter 5, you will be able to use, wear and eat polymers with a greater appreciation of the chemistry underpinning those diverse properties.

Answers to Questions

Question 1 The fraction by mass of gold in the Earth's crust is 4×10^{-9}.

So 4×10^{-9} g gold are contained in 1 g of the Earth's crust.

1 g gold is contained in $\dfrac{1}{4 \times 10^{-9}}$ g of the Earth's crust.

20 g gold is contained in $20 \times (\dfrac{1}{4 \times 10^{-9}})$ g of the Earth's crust.

The amount of earth that is required to extract 20 g gold is

$$20 \times (\frac{1}{4 \times 10^{-9}}) \text{ g} = 5 \times 10^9 \text{ g} = 5 \times 10^6 \text{ kg}.$$

Alternatively, this is 5 000 tonnes, where one tonne is equivalent to 1 000 kg. This amount of earth is sufficient to fill 250 twenty-tonne trucks (Figure 2.4).

Question 2 The empirical formulas are: CH_2, ZnS, CH_3, CaF_2 and $CuFeS_2$ respectively.

Question 3 Diamond has covalent bonds linking carbon atoms to form a three-dimensional structure. In contrast, the covalent bonds that link the carbon atoms in graphite result in a layer structure. Impurities can get between the layers and make it easier for these layers to slide over one another. It is for this reason that graphite can act as a lubricant as well as be used for the 'lead' in pencils. Diamond is one of the hardest substances known and it is its crystal structure that makes it much harder than graphite.

Question 4 *Caesium chloride:* There are eight chlorine atoms, each counting $\frac{1}{8}$ at the corners of the unit cell and a caesium atom in the body of the cell. The unit cell formula is $CsCl$ as is the empirical formula.

Zinc sulfide: There are four sulfur atoms within the body of the unit cell. Zinc atoms are in two environments: eight atoms at the corners of the cell each counting $\frac{1}{8}$ and a further six atoms at the centres of the cell faces each counting $\frac{1}{2}$, a total of four zinc atoms. The unit cell formula is Zn_4S_4 and the empirical formula is ZnS.

Calcium fluoride: The unit cell here is rather more complicated, with calcium atoms in both corner and face environments. There are eight calcium atoms at the unit cell corners each counting $\frac{1}{8}$ and six atoms at the centres of the faces each counting $\frac{1}{2}$, a total of four calcium atoms. In addition, there are eight fluorine atoms within the body of the cell to give a unit cell formula of Ca_4F_8 and the empirical formula is CaF_2.

Question 5 (a) 50 cm^3 of solution contains 2.3 g sodium chloride.

1 cm^3 of solution contains $\dfrac{2.3}{50}$ g sodium chloride.

1 000 cm^3 of solution contains $\dfrac{2.3}{50} \times 1\,000$ g sodium chloride.

The concentration of sodium chloride is 46 g l^{-1}.

(b) 700 cm^3 of solution contains 2 100 g sucrose.

1 cm^3 of solution contains $\dfrac{2\,100}{700}$ g sucrose.

1 000 cm^3 of solution contains $\dfrac{2\,100}{700} \times 1\,000$ g sucrose.

The concentration of the sucrose solution is 3 000 g l^{-1}.

(c) 0.5 cm^3 of solution contains 0.03 g sucrose.

1 cm^3 of solution contain $\dfrac{0.03}{0.5}$ g sucrose.

1 000 cm^3 of solution contains $\dfrac{0.03}{0.5} \times 1\,000$ g sucrose.

The concentration of the sucrose solution is 60 g l^{-1}.

Question 6 The balanced equations are:

(a) $FeO + H_2 = Fe + H_2O$ (the reaction was represented by a balanced equation in the question).

(b) $Fe_2O_3 + 3H_2 = 2Fe + 3H_2O$.

(c) $N_2 + 3H_2 = 2NH_3$.

(d) $CH_4 + 2Cl_2 = CH_2Cl_2 + 2HCl$ (the clue to part (d) is to spot that chlorine atoms appear in both the products, in CH_2Cl_2 and in HCl).

Question 7 Only (a) represents a balanced equation. Here, the numbers of atoms of each type are the same on each side and the total charges on each side are the same.

In (b), the numbers of atoms balance but the charges do not. There is no overall charge on the right but the charge on the left is +4 ($6 - 1 - 1 = 4$). The balanced equation is

$$BrO_3^- + 5Br^- + 6H^+ = 3Br_2 + 3H_2O.$$

The problem in (c) is that the numbers of atoms do not balance but the charges are balanced. The balanced equation is

$$2Ag^+ + Sn = 2Ag + Sn^{2+}$$

Question 8 (a) The mass of one mole of potassium bromide (KBr) is obtained by adding together the relative atomic mass of potassium in grams and the relative atomic mass of bromine in grams and is equal to $(39.1\,g + 79.9\,g) = 119\,g$. Of this, $39.1\,g$ is K^+ ions and $79.9\,g$ Br^- ions.

Our solution contains $\dfrac{20.0}{119}$ mol KBr, so the concentration of K^+ ions is

$$\frac{20.0}{119}\ mol\,l^{-1} = 0.168\ mol\,l^{-1}$$

and the concentration of Br^- ions is

$$\frac{20.0}{119}\ mol\,l^{-1} = 0.168\ mol\,l^{-1}$$

(b) The mass of one mole of sodium sulfide (Na_2S) is obtained by adding *twice* the relative atomic mass of sodium in grams to the relative atomic mass of sulfur in grams and is equal to $(2 \times 23.0\,g) + 32.1\,g = 78.1\,g$.

Our solution contains $\dfrac{20.0}{78.1}$ mol Na_2S, so the concentration of Na^+ ions is

$$2 \times \frac{20.0}{78.1}\ mol\,l^{-1} = 0.512\ mol\,l^{-1}$$

(note that there are two moles of sodium cations in one mole of sodium sulfide)

and the concentration of S^{2-} ions is

$$\frac{20.0}{78.1}\ mol\,l^{-1} = 0.256\ mol\,l^{-1}$$

(c) The mass of one mole of iron chloride ($FeCl_3$) is obtained by adding the relative atomic mass of iron in grams to three times the relative atomic mass of chlorine in grams and is equal to $55.8\,g + (3 \times 35.5\,g) = 162.3\,g$.

Our solution contains $\dfrac{20.0}{162.3}$ mol $FeCl_3$, so the concentration of Fe^{3+} ions is

$$\frac{20.0}{162.3}\ mol\,l^{-1} = 0.123\ mol\,l^{-1}$$

and the concentration of Cl^- ions is

$$3 \times \frac{20.0}{162.3}\ mol\,l^{-1} = 0.370\ mol\,l^{-1}$$

(Note that there are three moles of chloride anions in one mole of iron chloride.)

Question 9 Minerals are concentrated in the following major ways:
(i) Separation from molten magma by virtue of density differences.
(ii) Separation by differential crystallization from molten magma.
(iii) Dissolution in heated water and subsequent crystallization.
(iv) Concentration by water and wind.
Look back at Section 2.2 for more information.

Question 10 Table A1 shows the data represented in the pie chart in Figure 2.64.

The histogram is shown in Figure A1.

Table A1 Elemental analysis data in ascending order of abundance for an iron ore sample of mass 100 g.

Element	Mass/g
aluminium	1.0
magnesium	1.5
calcium	2.3
others	2.9
phosphorus	5.2
silicon	14.9
iron	31.1
oxygen	41.1

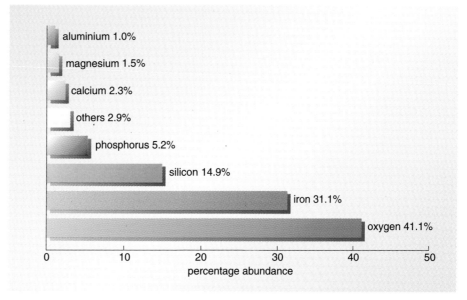

Figure A1
Histogram showing the elemental analysis data in descending order of abundance for an iron ore sample.

Question 11 There is just one calcium cation and it is within the body of the unit cell (Figure 2.65). There are eight titanium ions at the corners of the unit cell each counting $\frac{1}{8}$ to give a total of one titanium ion. There are twelve oxygen anions on the edges of the unit cell each counting $\frac{1}{4}$ to give three oxygen anions. The unit cell formula of perovskite is $CaTiO_3$ and the empirical formula is also $CaTiO_3$.

Overall, the compound is uncharged so the total of the positive charges must be equal to the total of the negative charges for the ions in the unit cell. The total of the charges for the negative ions is $3 \times (-2) = -6$. The total of the charges for the positive ions is $(2 + n)$ where n is the charge on the titanium ion. So

$$(2 + n) = 6$$

By subtracting 2 from both sides of the equation, we get

$$n = 6 - 2 = 4$$

The charge on the titanium ion is + 4.

Question 12 The relative atomic mass of magnesium is 24.3 and that of iodine 127.

One mole of magnesium iodide has a mass of $(24.3 + (2 \times 127))$ g = 278.3 g.

The solution contains $\dfrac{27.8}{278.3}$ = 0.100 mol of magnesium iodide in a volume of 105 cm^3.

1 cm^3 of the solution contains $\dfrac{0.100}{105}$ mol of magnesium iodide.

1 litre of the solution contains $1\,000 \times \left(\dfrac{0.100}{105}\right)$ mol of magnesium iodide.

In this solution, there are two iodide anions for each magnesium cation, so the concentration of the solution in terms of the iodide ion I$^-$ is

$$2 \times 1\,000 \times \left(\frac{0.100}{105}\right) \text{mol l}^{-1} = 1.90 \text{ mol l}^{-1}.$$

Question 13 Concrete is a relatively cheap building medium which can be manufactured in bulk. It is possible to mix the concrete at a depot and transport it to a building site where it can be poured and moulded into almost any shape. Concrete is particularly strong under compression and structures are designed to maximize this quality of concrete. It has low tensile strength although this can be improved with steel reinforcement. However, care must be taken to minimize water penetration of the concrete and rusting of the reinforcement.

Question 14 Glass does not conduct heat well and this renders it prone to cracking when subjected to rapid temperature changes. This problem can be overcome by producing a glass with low expansion properties (by including boron in the glass structure) and by using very thin glass. Both approaches are used in the Dewar flask but the flask does have to protected against mechanical shock.

Question 15 What is required is a celadon glaze. Particles of pure silica should be used and firing should allow partial melting. Light will be able to penetrate to the vessel surface through the transparent melted silica. The unmelted particles will cause the glaze surface to be uneven and there will be irregular reflections. The colour spots can be created with particles of a cobalt compound that will not melt at the firing temperature.

Question 16 (a) The metals with the highest electrical conductivity at 25 °C are silver, copper and gold. The high cost of silver and gold limits their use in all but the most specialized of electrical components. Pure copper is therefore the metal for general electrical use.

(b) Aluminium has the advantage of low density which is essential for aeronautical engineering. It is also strong and has good resistance to corrosion but it is not particularly cheap.

(c) Tin has quite a low melting temperature for a metal. Iron objects could be dipped into liquid tin without the iron melting.

(d) The short answer is none. Gold has a low abundance and is certainly very costly. Lead has quite a low abundance and yet its cost is relatively low. Aluminium is very abundant but it commands a significant price in the market. The cost of a metal depends on a number of factors which include its concentration and distribution in the Earth's crust, methods available for extraction, market demand and sometimes political considerations.

Question 17 Formula mass for covellite (CuS) = 63.5 g + 32.1 g = 95.6 g

The proportion of copper by mass in covellite is = $\dfrac{63.5\,\text{g}}{95.6\,\text{g}}$ = 0.664

Mass % of copper = 0.664 × 100 = 66.4%

Question 18 FeO: Formula mass for FeO = 55.8 g + 16.0 g = 71.8 g

Fraction of iron = $\dfrac{55.8\,\text{g}}{71.8\,\text{g}}$ = 0.777

Mass % of iron = 0.777 × 100 = 77.7%

Fe_2O_3: Formula mass for Fe_2O_3 = (2 × 55.8 g) + (3 × 16.0 g) = 159.6 g

Fraction of iron = $\dfrac{111.6\,\text{g}}{159.6\,\text{g}}$ = 0.699

Mass % of iron = 0.699 × 100 = 69.9%

Fe_3O_4: Formula mass for Fe_3O_4 = (3 × 55.8 g) + (4 × 16.0 g) = 231.4 g

Fraction of iron = $\dfrac{167.4\,\text{g}}{231.4\,\text{g}}$ = 0.723

Mass % of iron = 0.723 × 100 = 72.3%

Question 19 Remember that one mole of anything contains 6.02×10^{23} of the specified entities.

(a) 1.00 mol of chlorine atoms contains 6.02×10^{23} atoms. The mass is (1.00 × 35.5) g = 35.5 g.

(b) 1.00 mol of hydrogen molecules contains 6.02×10^{23} molecules. The mass is (1.00 × 2.02) g = 2.02 g. (Remember that the molecular formula for the hydrogen molecule is H_2.)

(c) 0.50 mol of water molecules contains 0.50 × (6.02×10^{23}) = 3.01×10^{23} molecules. The mass is (0.50 × (2.02 + 16.0)) g = 9.01 g.

(d) 1.50 mol of oxygen molecules contains 1.50 × (6.02×10^{23}) = 9.03 × 10^{23} oxygen molecules. The mass is (1.50 × 32.0) g = 48.0 g.

(e) 2.00 mol of nitrate anions contains 2.00 × (6.02×10^{23}) = 12.04×10^{23} nitrate anions. The mass is (2.00 × (14.0 + 48.0)) g = 124 g.

Question 20 (a) $CuS(s) + O_2(g) = Cu(s) + SO_2(g)$.

The equation tells us that one mole of copper sulfide produces one mole of sulfur dioxide.

The formula mass of CuS is $63.5\ g + 32.1\ g = 95.6\ g$.

Relative molecular mass of SO_2 is $32.1\ g + (2 \times 16.0\ g) = 64.1\ g$.

So $95.6\ g\ CuS$ produces $64.1\ g\ SO_2$.

So $1.00\ g\ CuS$ produces $\dfrac{64.1}{95.6}\ g\ SO_2$.

So $50.0\ kg\ CuS$ produces $\dfrac{64.1}{95.6} \times 50.0\ kg = 33.5\ kg\ SO_2$.

(b) The equation representing the reaction is

$$Fe_3O_4(s) + 2C(s) = 3Fe(s) + 2CO_2(g).$$

From the equation, we can see that one mole of Fe_3O_4 reacts with two moles of carbon.

The formula mass of Fe_3O_4 is $(3 \times 55.8\ g) + (4 \times 16.0\ g) = 231.4\ g$.

So $1\ g\ Fe_3O_4$ reacts with $\left(\dfrac{2\ \times\ 12.0}{231.4}\right)\ g$ of carbon.

So $1\ kg\ Fe_3O_4$ reacts with $\left(\dfrac{2\ \times\ 12.0}{231.4}\right)\ kg$ of carbon.

So $1\ 000\ kg\ Fe_3O_4$ reacts with $1\ 000 \times \left(\dfrac{2\ \times\ 12.0}{231.4}\right)\ kg = 104\ kg$ of carbon.

Question 21 $100\ g$ of pig iron would contain $95\ g$ iron and $5\ g$ carbon.

$95\ g$ iron is equivalent to $\dfrac{95}{55.8}$ mol iron.

There are $\dfrac{95}{55.8} \times (6.02 \times 10^{23})$ atoms in $95\ g$ iron.

$5\ g$ carbon is equivalent to $\dfrac{5}{12.0}$ mol carbon.

There are $\dfrac{5}{12.0} \times (6.02 \times 10^{23})$ atoms in $5\ g$ carbon.

The ratio of the number of iron atoms to the number of carbon atoms is given by the fraction

$$\left(\dfrac{\dfrac{95}{55.8} \times\ 6.02\ \times\ 10^{23}}{\dfrac{5}{12.0} \times\ 6.02\ \times\ 10^{23}}\right) = \left(\dfrac{12.0}{5}\ \times\ \dfrac{95}{55.8}\right)$$

which is very close to four.

Although carbon represents only one-twentieth of the mass of pig iron, it represents about one-quarter of the number of atoms.

Question 22   The critical information required to answer this question is the fact that the concentration of $H^+(aq)$ multiplied by the concentration of $OH^-(aq)$ is equal to $1.0 \times 10^{-14} \text{ mol}^2 \text{ l}^{-2}$.

(a) $1.0 \times 10^{-14} \text{ mol l}^{-1} \times [OH^-(aq)] = 1.0 \times 10^{-14} \text{ mol}^2 \text{ l}^{-2}$

$$[OH^-(aq)] = \frac{1.0 \times 10^{-14} \text{ mol}^2 \text{ l}^{-2}}{1.0 \times 10^{-14} \text{ mol l}^{-1}} = 1.0 \text{ mol l}^{-1}$$

(b) $1.0 \times 10^{-4} \text{ mol l}^{-1} \times [OH^-(aq)] = 1.0 \times 10^{-14} \text{ mol}^2 \text{ l}^{-2}$

$$[OH^-(aq)] = \frac{1.0 \times 10^{-14} \text{ mol}^2 \text{ l}^{-2}}{1.0 \times 10^{-4} \text{ mol l}^{-1}} = 1.0 \times 10^{-10} \text{ mol l}^{-1}$$

(c) $1.0 \times 10^{-7} \text{ mol l}^{-1} \times [OH^-(aq)] = 1.0 \times 10^{-14} \text{ mol}^2 \text{ l}^{-2}$

$$\frac{1.0 \times 10^{-14} \text{ mol}^2 \text{ l}^{-2}}{1.0 \times 10^{-7} \text{ mol l}^{-1}} = 1.0 \times 10^{-7} \text{ mol l}^{-1}$$

(d) $1.0 \text{ mol l}^{-1} \times [OH^-(aq)] = 1.0 \times 10^{-14} \text{ mol}^2 \text{ l}^{-2}$

$$[OH^-(aq)] = 1.0 \times 10^{-14} \text{ mol l}^{-1}$$

Question 23   The pH values are obtained by taking the exponent and changing the sign.

(a) pH = 14, (b) pH = 4, (c) pH = 7, (d) pH = 0. (Remember from Book 1 that $10^0 = 1$.)

Question 24   (a) Copper is more easily oxidized than is silver so the following reaction should occur:

$$Cu(s) + 2Ag^+(aq) = Cu^{2+}(aq) + 2Ag(s).$$

(b) Tin is more easily oxidized than is copper so the following reaction should occur:

$$Sn(s) + Cu^{2+}(aq) = Sn^{2+}(aq) + Cu(s).$$

(c) Copper lies above silver in the activity series. Silver should not be oxidized by $Cu^{2+}(aq)$ ions.

Question 25   (a) Iron is above tin in the activity series and therefore more easily oxidized than tin. Iron should be oxidized to $Fe^{2+}(aq)$ and $Sn^{2+}(aq)$ reduced to elemental tin.

$$Fe(s) + Sn^{2+}(aq) = Fe^{2+}(aq) + Sn(s).$$

(b) Aluminium is above iron in the table so it should be oxidized.

$$2Al(s) + 3Fe^{2+}(aq) = 2Al^{3+}(aq) + 3Fe(s)$$

(c) Zinc is below magnesium in the table so no reaction should occur.

Question 26   The formula mass of Fe_2O_3 is equal to $(2 \times 55.8 \text{ g}) + (3 \times 16.0 \text{ g}) = 159.6 \text{ g}$. The relative atomic mass of C is equal to 12.0. The equation for the reaction is

$$2Fe_2O_3 + 3C = 4Fe + 3CO_2$$

and from the equation we can say that two moles of Fe_2O_3 react with three moles of carbon atoms.

So 1 g Fe_2O_3 reacts with $\dfrac{3 \times 12.0}{2 \times 159.6}$ g carbon.

So 1 tonne Fe_2O_3 reacts with $\dfrac{3 \times 12.0}{2 \times 159.6}$ tonne carbon.

So 0.113 tonne of carbon is required to react with 1 tonne Fe_2O_3.

Question 27 The balanced equations are:

(a) $Fe(s) + 2H^+(aq) = Fe^{2+}(aq) + H_2(aq)$

(b) $Ca(OH)_2(s) + 2H^+(aq) = Ca^{2+}(aq) + 2H_2O(l)$

(c) $2CrO_4^{2-}(aq) + 2H^+(aq) = Cr_2O_7^{2-}(aq) + H_2O(l)$

Question 28 The relative molecular mass of nitric acid is 63.01.

63.01 g nitric acid contains 1.01 g hydrogen.

200 cm³ solution will then contain 1.01 g hydrogen cations.

1 cm³ solution will then contain $\dfrac{1.01}{200}$ g hydrogen cations.

1 000 cm³ solution will then contain $\dfrac{1.01}{200} \times 1\,000$ g hydrogen cations.

The concentration of the solution in hydrogen cations is 5.05 g l⁻¹. Note: it is important to specify the species to which the concentration data refer.

Question 29 This is a reaction between a solution of an acid and a base and can be simply represented by a reaction between hydrogen and hydroxide ions:

$H^+(aq) + OH^-(aq) = H_2O(l)$

You may choose to include all the ions present in nitric acid and in sodium hydroxide as in the equation below. This is not incorrect but it is unnecessary to show the species that do not actually take part in the reaction:

$H^+(aq) + NO_3^-(aq) + Na^+(aq) + OH^-(aq) = H_2O(l) + NO_3^-(aq) + Na^+(aq)$

Question 30 To balance such an equation looks to be rather tricky but let's just think for a moment. The reaction involves hydrogen ions and these will react with the hydroxide ions in magnesium hydroxide. The reaction can be represented by

$H^+(aq) + Mg(OH)_2(s) \;/=/\; Mg^{2+}(aq) + H_2O(l)$.

The balanced equation is

$2H^+(aq) + Mg(OH)_2(s) = Mg^{2+}(aq) + 2H_2O(l)$.

Note that the equation balances both in terms of the numbers of different types of atoms on each side and the total charges on each side are the same.

The above equation represents overall what happens in the reaction. If we were to include all the species that are present, you probably will agree that the result below is much more complicated:

$$2H_3PO_4(aq) + 3Mg(OH)_2(s) = 3Mg^{2+}(aq) + 2PO_4^{3-}(aq) + 6H_2O(l)$$

Question 31 This question is designed to emphasize the differences between these similar-sounding terms.

An isotope of an element is an atom of that element which has a specified number of neutrons in the nucleus. For example, carbon-12 has a nucleus with six protons and six neutrons. (There are also six electrons around the nucleus in the neutral atom.) The isotope carbon-13 has again six protons in its nucleus but seven neutrons. Both carbon-12 and carbon-13 are isotopes of carbon.

Allotropes are different structural forms of the same element. Both diamond and graphite contain only carbon atoms but the atoms are bonded together to give different structures which are shown up in the distinctive physical properties of diamond and graphite.

The term isomer refers to molecules which have the same molecular formula but have different structures. Figure A2 shows ball-and-stick models of two molecules of molecular formula C_6H_{14}. They each have six carbon atoms and fourteen hydrogen atoms but have different properties.

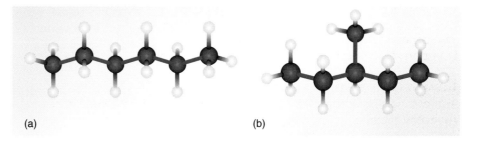

Figure A2
Hexane (a) and methylpentane (b) are isomers.

(a)   (b)

Question 32 You should recall that the names of many polymers are concocted simply by placing 'poly' before the monomer name. (a) Thus, poly(tetrafluoroethene) or PTFE is made from tetrafluoroethene. (b) The systematic name of Acrilan indicates that propenenitrile is the monomer used to make this fibre. Therefore, the question asks for the common name of the monomer for Acrilan. Table 5.1 lists the monomer as acrylonitrile. In fact, acrylonitrile is the common and alternative name for propenenitrile.

Question 33 Octane has two CH_3 groups linked by six CH_2 groups, so its shorthand form is $CH_3(CH_2)_6CH_3$.

Question 34 The systematic name for polythene is polyethene or, strictly, poly(ethene) so its monomer is ethene. In Book 1 you learned that ethene has the formula C_2H_4 and has two carbon-to-carbon bonds, that is, it contains a double bond. It is a flat and inflexible molecule. You built a model of ethene and hence witnessed its shape.

Question 35 Both acrylonitrile and propene (or propylene) will react by addition polymerization because they have carbon-to-carbon double bonds. They provide the important polymers polyacrylonitrile and polypropylene, respectively. The monomer units will be $-CH_2-CHCN-$ and $-CH_2-CHCH_3-$. Propane does not have a double bond so it cannot participate in addition reactions. (Neither does it have functional groups so it will not undergo condensation polymerization.)

Question 36 In addition polymerization, the monomers simply add to each other and do not produce a by-product of any sort. Condensation polymerization has water as a by-product.

Monomers for addition polymerization need to contain only a carbon-to-carbon double bond. A wide range of other functional groups can be involved in condensation polymerization.

A more subtle point is that the 'backbone' of an addition polymer is simply a string of carbon atoms, originally the carbons atoms that were held by the double bond in the monomer. We have seen that connecting groups in condensation polymers contain other atoms: nitrogen in polyamides and oxygen in polyesters. The backbone of the polymer in Figure 5.28 is more complicated than a chain of carbon atoms. It contains nitrogen in its connecting groups and it is a polyamide. Therefore, it was produced by condensation polymerization.

Question 37 We may expect that in polymers of this type the chains will be held firmly together once they have been made, and so the polymer will not soften on heating. They are called thermosetting polymers.

Question 38 See Table A2.

Table A2 Completed Table 5.5, for use with answer to Question 38.

Term	Terylene	Polyethene	PVC	Polyisoprene	Nylon	Bakelite
condensation polymer	✓				✓	✓
cross-linked polymer						✓
thermosetting plastic						✓
polyamide					✓	
thermoplastic polymer	✓	✓	✓	✓	✓	
addition polymer		✓	✓	✓		
polyester	✓					

Question 39 Thermosetting polymers cannot be softened by heat, and so the 'remelting' routes are not open to them. The most effective recycling opportunities involve grinding the products and using the material as filler in other mixtures. However, as the fibre length will be shortened (e.g. recycled paper), the product may have lower quality than that of virgin materials.

Question 40 In examining the relative environmental benefits of using paper or polystyrene foam cups, we can first compare the amount of raw material and energy used in their manufacture. From the table, it is clear that the paper cup requires much more raw material and uses more energy. The organic chemicals needed in the paper-making process, which are also reflected in the amounts in the water effluent, provide a potential for considerable environmental impact compared with the only major release from polystyrene manufacture – the pentane used to 'blow' the foam.

Reuse and recycling are also easier for the polystyrene cup, which also offers greater potential in terms of its heat content for recovery on incineration. However, because landfill is a major disposal route in the UK, the biodegradability potential favours the paper cup, although recent studies have found that even 'biodegradable' materials like paper may remain intact in a landfill site over very long periods.

On the basis of this analysis, the polystyrene cups appear to have the environmental advantage. However, when analysing data of this type you should always look critically at the information provided. Typical omissions of life cycle analyses have been the failure to include an impact from the transport of raw materials and of products, as well as the environmental impact of the capital equipment used in the manufacturing process and, indeed, the vehicles for transport. Life cycle analysis is still a relatively new technique, requiring standardization in its approach. Until that is done, comparisons of the type in this question must be made carefully.

Question 41 Various additives have been mentioned throughout this Chapter, and include those listed in Table A3. To these you may add fibres for reinforcement.

Table A3 For use with answer to Question 41.

Additive	Function
stabilizers or fillers	increase strength, reduce costs, impart fire resistance, and prevent degradation by sunlight
plasticizers	increase flexibility
colouring agents	impart colour
flame retardants	reduce flammability
cross-linking agents	promote cross-linking

Acknowledgements

Grateful acknowledgement is made to the following sources for permission to reproduce material in this part of Book 2:

Figure 1.3 NASA; *Figure 1.7* Staatliche Museen zu Berlin, Gemaldegalerie; *Figures 2.5, 4.5 and 5.3* Natural History Museum Picture Library; *Figure 2.20 (top)* De Beers Consolidated Mines; *Figures 2.35, 4.14 and 4.51* Copper Development Association, Potters Bar, Herts; *Figure 2.45* Bath Abbey; *Figure 2.54* BP Chemicals; *Figures 2.60 and 5.13* Ardea Ltd; *Figure 3.1* Aerial Archaeology Publications: photo by Derek A. Edwards; *Figure 3.5* Yorkshire Dales National Park; *Figures 3.10, 3.21, 4.2 and 4.3* British Museum; *Figures 3.14, 3.23, 3.24, 3.27 and 3.28* City Museum and Art Gallery, Hanley, Stoke-on-Trent; *Figure 3.1* BNFL; *Figure 4.6* Peter Newark's Western Americana; *Figure 4.14* IMI, Yorkshire; *Figure 4.19* E. G. Holland, Cicerone Press, Cumbria; *Figure 4.22* courtesy of Billiton International Metals B.V.; *Figure 4.23* courtesy of Wardown Park Museum, Luton, Bedfordshire; *Figure 4.26* Michael Holford; *Figure 4.28* Philip Chinnery; *Figure 4.29* courtesy of Weald and Downland Open Air Museum, Singleton, West Sussex; *Figures 4.32 and 4.33* courtesy of British Steel, Irthlingborough; *Figure 4.36* courtesy of the Chemical Society Library; *Figure 4.48* courtesy of the Galvanisers' Association; *Figure 4.53* Science Photo Library; *Figure 5.15* British Textile Technical Group; *Figure 5.16* Heather Angel/Biophotos; *Figure 5.17* Science Life Library (1966) *Giant Molecules* © 1966 Time–Life Books Inc.; *Figures 5.24 and 5.26* courtesy of Hagley Museum and Library; *Figure 5.35* H. Allen (1949) *The House of Goodyear: Fifty Years of Men and Industry,* Corday & Gross, Cleveland.

This part of Book 2 explores the practical consequences of the energy changes that accompany chemical processes – in the laboratory, in the home, nationally and globally – and, in so doing, establishes some of the underlying chemical principles.

PART 2 ENERGY

Prepared for the Course Team by Michael Mortimer

Contents

Chapter 1
Introduction

Energy, its availability and our ability to manipulate and harness it for practical purposes, has had a major influence on the growth of modern industrialized society. Many of the things we take for granted as essential to our standard of living – heat and electricity for our homes, convenient transport systems, plentiful food supplies and a wealth of manufactured goods – all depend upon a continuous supply of energy. On a day-to-day basis, energy is a commodity that we cannot do without; it has that vital capacity which allows us to do things.

At the time of writing, most daily energy needs in the developed world are provided, in one way or another, by the fossil fuels; i.e. coal, oil and natural gas. Other energy sources, notably nuclear, hydroelectric, solar, wind and wave power, make smaller contributions; and food, of course, provides the energy necessary for life itself. However, to highlight the dependence on fossil fuels is to introduce an uncomfortable reality: one day, they will all be exhausted. This means it is imperative to use what remains as efficiently as possible and, just as importantly, it is vital to develop alternative energy sources for the future.

The current demands for energy, driven by ever-increasing pressures for economic growth, must also be satisfied with minimum damage to the environment. On a global scale, the natural energy systems on which the heat balance and climate of the Earth depend must not be distorted; and, on a local scale, pollution due to emissions from energy-producing processes must be restricted. It is not surprising that energy issues are often controversial and rarely straightforward.

Chemistry and chemists are very much concerned with all aspects of energy provision and use. It is this relationship, and an exploration of it, that provides a main theme for this part of Book 2. A major concern will be with what we as consumers should know and understand about the links between chemistry and energy. This will take us on a journey from the laboratory, into the home and, finally, to discuss energy matters on a national and global scale; a constantly recurring theme will be to seek explanations at the molecular level. Along the way, we shall establish some of the underlying chemical principles and we shall see also that the concept of energy itself plays a central role in chemical discussions. The chemical principles that we uncover provide the stepping stones that allow the story to move forwards. They also provide the necessary background that will allow you to read and critically assess articles that touch on chemical aspects of energy issues. The montage overleaf hints at some of the areas to be covered in this part of Book 2.

World energy consumption grew by 0.2% in 1993, so bringing total demand almost back to 1990's all-time high. The virtually flat profile since 1990 marks a break from the pattern of steady increases between 1983 and 1990, when the growth rate averaged 2.8% a year.
BP Statistical Review of World Energy, June 1994

The Sun is the source from which almost all energy on the Earth is derived.

Figure 1.1

Chapter 2
Energetic chemical reactions

2.1 More than they may seem

So far in this Book you have met a selection of chemical reactions and have seen how they can be described in terms of the shorthand notation of balanced chemical equations. These equations provide a lot of information. They identify the reactants and products and give their respective formulas. They also indicate the relative numbers of each reactant molecule that is involved in the reaction and how many molecules of each product are formed.

■ The compound aluminium bromide ($AlBr_3$), which is a white solid at room temperature, can be prepared by the reaction between aluminium (Al) metal and bromine (Br_2) liquid. These materials are shown in Figure 2.1. In unbalanced form, the chemical equation that describes the reaction is

$Al(s) + Br_2(l) \,/=/\, AlBr_3(s)$

What is the form of the balanced chemical equation?

■ Using the approach described in the first part of this Book, the balanced chemical equation is

$2Al(s) + 3Br_2(l) = 2AlBr_3(s)$

Figure 2.1
Aluminium (left), bromine (centre) and aluminium bromide (right). As can be seen in the photograph, bromine is a dark red liquid with a fairly dense vapour above its surface; for this reason it is sometimes referred to as a 'fuming liquid'. In the liquid (and vapour), it is known that the bromine atoms, chemical symbol Br, are chemically bound together in pairs to form bromine molecules. Bromine liquid is thus written as $Br_2(l)$.

Balanced chemical equations serve an important purpose in chemistry but they do convey a rather 'static' image; that is, they give no indication of how reactions occur in practice. Figure 2.2 illustrates this point dramatically. It shows a snapshot of the situation just a few seconds after shreds of aluminium metal are added to a beaker containing a small amount of bromine liquid.

How would you describe what happens in Figure 2.2?

Figure 2.2
A few seconds after shreds of aluminium are added to bromine liquid.

A very vigorous reaction obviously occurs; the mixture 'sets on fire' and clouds of vapour appear. The situation is very different from the tranquil scene in Figure 2.1. The violent nature of the reaction is certainly not conveyed by the simple statement that aluminium metal reacts with bromine liquid to produce a white powder called aluminium bromide. You may also have suggested, on the basis of seeing the flames in the beaker, that the reaction gives out heat as it takes place. This is an important observation. Incidentally, it also accounts for the clouds of vapour, since any unreacted liquid bromine would be quickly heated to its boiling temperature (a relatively low 59 °C) and so be turned completely into vapour. To make matters more complicated, the product, aluminium bromide, also vaporizes to some extent to give white fumes.

The reaction we've just described is by no means one that is used routinely in the chemical laboratory. It does, however, illustrate quite vividly an important property of many chemical reactions; that is, when they take place they produce heat and so raise the temperature of their local surroundings.

The production of heat by a chemical reaction is, in fact, quite a familiar process. For example, you may use natural gas to keep your home warm or for cooking. As you may recall, the major component of natural gas is methane with molecular formula CH_4. Normally, we think of burning the gas in 'air' but strictly it is the oxygen in the air that is the key ingredient. The balanced chemical equation for the reaction is

$$CH_4(g) + 2O_2(g) = CO_2(g) + 2H_2O(g)$$

and of course, as experience tells us, heat is also produced (Figure 2.3). The water that is formed in the reaction is in the form of a vapour and could be detected, for example by allowing it to condense on a cold object that is held above the flame. Whenever an element, or compound, burns in a plentiful supply of oxygen, the overall process is referred to as **combustion**.

> **Question 1** Propane, which has molecular formula C_3H_8, is a gas under normal conditions. Commercially, it is sold in metal bottles in which it is contained at a high pressure – for this reason it is often called 'bottled gas'. Mixed with air and ignited, it burns with a hot flame. See if you can develop a balanced chemical equation to describe what happens when propane gas burns in air. You can assume, as in the case of combustion of methane, that the only products are carbon dioxide and water vapour.

Firework displays provide further examples of production of heat by chemical reactions, although you may not have thought about them in

Figure 2.3
A natural gas hob in use.

Figure 2.4
Part of the firework display for the opening ceremony of the 1992 Olympic Games at Barcelona.

such terms. The firework, or pyrotechnic, effects that have become such a familiar part of major celebrations are becoming increasingly sophisticated, but the displays you have probably had at some time in your own garden involve the same basic principles.

The detailed recipe for a firework mixture is often a secret, but in principle there are just two main ingredients, both of which are solids. One of these is the fuel which combines with oxygen; e.g. it could be a powder of aluminium or magnesium, charcoal, a suitable organic compound, or a mixture of these. The other is the oxidizer which has oxygen chemically bound within its structure. Typically, it could be a simple metal nitrate such as sodium nitrate, $NaNO_3$(s). It is the oxidizer which provides the oxygen source; in other words, the mixture has its own 'in-built' oxygen supply and does not rely on oxygen from the air. Igniting the blue touch paper results in the mixture inside the firework being heated and this starts the chemical reaction in which the fuel is oxidized. As you should recall, *oxidation* can simply be interpreted as meaning 'combining with oxygen' and so if the fuel was magnesium metal it would be converted to magnesium oxide, MgO(s). The heat produced by the oxidation reaction tends to melt the firework mixture with the result that there is a far more intimate mixture of fuel and oxidizer than when both ingredients were solids. This is just one of the factors that makes the oxidation reaction proceed with increasing speed.

The bright sparks propelled from fireworks are usually burning metal particles which continue to burn by drawing on oxygen from the air. The bigger the particles, the longer the sparks last. Magnesium particles reach temperatures of over 2 500 °C and the magnesium oxide formed glows white-hot. Iron particles reach much lower temperatures and produce dimmer, gold-coloured sparks. Different coloured flames are achieved by adding different metal compounds to the firework mixture; calcium salts for orange–red, strontium salts for crimson, and barium salts for yellow–green flames.

Box 2.1 A sophisticated firework

The design of a firework (Figure 2.5) determines its visual effect. The shell, which is 7 cm to 30 cm in diameter, is launched from a mortar tube. To effect its launch, the propellant charge in the bottom of the shell is ignited by a quick-burning fuse and the whole shell is thrown a few hundred metres into the air. A time-delay fuse also begins to burn when the shell is set off; some seconds later, when the shell is far above the ground, the bursting charge is ignited. When the charge explodes, it in turn ignites the numerous stars and scatters them in a symmetrical pattern.

Depending on the size and chemical composition of the stars, there can be bright coloured flashes, extended trails or explosions. Black powder (the original gunpowder) is used in both the propellant and bursting charges. The Chinese developed the basic formula more than 1 000 years ago and this has remained substantially unchanged over the centuries. It is close to being an ideal pyrotechnic substance because it consists of abundant, inexpensive chemicals that are relatively non-toxic. However, it is a *dangerous* substance because it is easily ignited by a moderate jolt of energy, such as a spark.

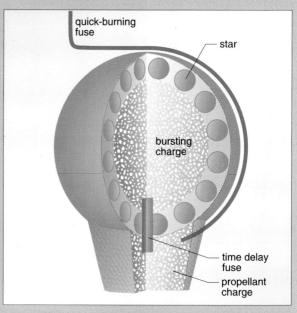

Figure 2.5
The structure of a Japanese-style 'chrysanthemum shell' firework.

Question 2 Carbon dioxide is used in one type of fire extinguisher. The effect of the gas is to smother the flames and deprive whatever is burning of oxygen; without oxygen, most fires will die out. However, a carbon dioxide fire extinguisher would be of no use for a magnesium fire because this metal will burn in carbon dioxide. The products formed are magnesium oxide and carbon. See if you can develop a balanced chemical equation to describe the reaction. How would you describe it in chemical terms?

You may have gained the impression so far that all chemical reactions that produce heat burst into flames automatically! This is certainly not the case; indeed, the chemistry laboratory would be a very unsafe place if it were. For now, we can simply note that different chemical reactions produce different quantities of heat and in many cases the output is quite modest.

To take the discussion further, it's useful to change direction a little and consider whether or not a heat change occurs when a simple ionic compound dissolves in water. We shall choose potassium nitrate, $KNO_3(s)$, as an example.

▨   What is the equation that represents the process of dissolving potassium nitrate, KNO_3, in water?

▨   The equation for the dissolution is

$$KNO_3(s) = K^+(aq) + NO_3^-(aq)$$

To find out whether there is a heat change, you can carry out your own experiment.

Experiment 2.15 Detecting a heat change

In your experiment, you found that dissolving potassium nitrate in water produced a solution which was colder than the water you started with; i.e. there was a drop in temperature. This may have surprised you. In effect, the dissolution process 'took in' heat from its local surroundings, including the solution itself, which was left colder than it was at the outset. Of course, if you had waited long enough, the solution would have 'absorbed' heat from its surroundings and so returned to the temperature of the water at the start of the experiment. It is in this sense, *that is by reference to the surroundings*, that we can state that the dissolution process '*absorbs heat*'. By the same token, the combustion reactions we considered earlier '*release heat*' to the surroundings.

Dissolving potassium nitrate in water is clearly a simple process to observe experimentally; hence its choice as an example. However, the observation that heat is absorbed is very important. This property is not in any way special and, in fact, it turns out that many chemical reactions behave in a similar manner.

There may be a temptation for you to conclude from your experiment that heat is always absorbed when an ionic compound dissolves in water. This should be resisted! If a number of ionic compounds had been at your disposal, and you had carried out similar experiments, you would have observed in some cases a temperature rise in the solution and in others a temperature fall. Whether heat is produced or absorbed when an ionic compound dissolves in water depends very much on the nature of the compound itself.

Figure 2.6
A very cold reaction mixture and a frosted beaker results from a reaction that 'absorbs heat'.

Reactions that absorb heat as they progress are usually far less dramatic than those that burst into flames. However, one particular reaction is often quoted because of its visual effect. It involves the reaction between two solids. (These are ammonium thiocyanate and barium hydroxide octahydrate but the detailed chemistry need not concern us.) As the reaction progresses, it absorbs a great deal of heat from its local surroundings, with the result, as shown in Figure 2.6, that the reaction mixture becomes very cold and frost forms on the outside of the beaker in which it takes place.

So, what have we learnt about chemical reactions so far? How would you summarize the preceding discussion?

The key point is that chemical reactions can release heat to, or absorb heat from, their surroundings. As you will see, an alternative way of expressing this is to state that *chemical reactions involve energy changes*. A major part of the remainder of this part of Book 2 will be devoted to understanding why these energy changes occur and, in turn, how they are used practically. Before moving on, however, mention must be made of one other very important practical feature of a chemical reaction that cannot be inferred from the form of its balanced equation. Can you suggest what this is?

The answer, to use an everyday expression, is the 'speed' of the reaction, that is how quickly the reactants are consumed or the products are formed. The reaction between aluminium metal and bromine liquid started almost immediately the two materials came into contact and then proceeded very rapidly – an observation you certainly would have made if you had been present when the photograph was taken. In marked contrast, as you know from experience, if a gas tap is turned on, the escaping gas doesn't immediately burst into flames. Indeed, a mixture of natural gas and air will essentially remain unchanged for as long as you wish to wait. To start, or initiate, the reaction, a spark or flame is required; without this stimulus we can view the reaction as occurring so slowly that no change is ever observed. The balanced chemical equation for the combustion of methane gas contains none of this information.

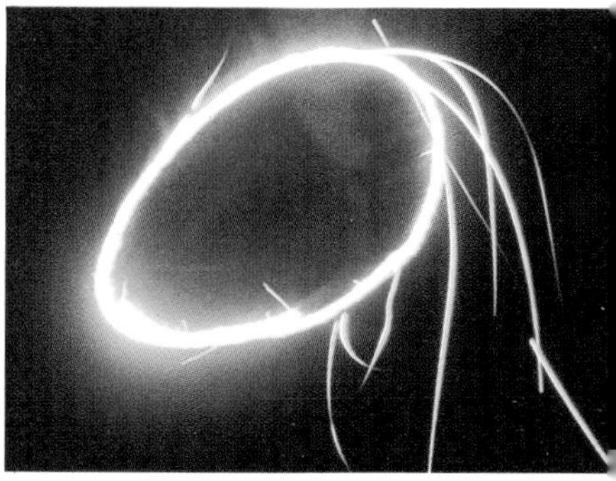

The speed of a chemical reaction, depending on the materials involved and circumstances, can vary from being explosively fast to tediously slow. This whole area of study of the speed of reactions is referred to as **chemical kinetics**. It has many important applications, for example in the chemical industry a knowledge of the factors that determine the speed of a reaction is essential to the design of efficient plants and processing methods. It also plays a role in understanding the chemistry that occurs in the human body, and it helps in characterizing the reactions that are going on in the Earth's atmosphere. The design of sparklers also involves chemical kinetics, as described in the caption to Figure 2.7.

Figure 2.7
Red sparklers are made from a mix containing several ingredients. A strontium compound is used to give the red colour and iron filings the sparkle. The oxidizer is a metal nitrate. To ensure that the sparkler lasts a reasonable length of time, the fuel must not burn too rapidly. Aluminium, rather than magnesium, powder is the choice in this case.

2.2 Candles and flames

Flames are an integral part of burning and combustion, and clearly provide strong evidence that chemical reactions may involve large energy changes. But why is a flame formed? What are its properties? This section digresses a little to provide some answers to these questions. In the process, we discover quite a lot about the way in which many chemical reactions actually occur.

The words accompanying the photograph of the burning candle in Figure 2.8 are taken from a very famous series of Christmas lectures for young people given by Michael Faraday (1791–1867) when he was director of the Royal Institution in London in the mid-19th century*. Faraday carried out pioneering experiments in both physics and chemistry during his time at the Institution and is justly recognized today as one of the great scientists of his era. Characteristically, the theme for his lectures was selected on the basis of the simple and familiar; hence the burning candle. This commonplace phenomenon provided the reference point to describe what was then known about the principles of chemistry. The Royal Institution Christmas lectures still continue and are now televised, commanding an audience far larger than Faraday could ever have imagined.

The candle still remains familiar, although nowadays it is largely confined to dinner tables and birthday cakes. The flame, which is self-sustaining, always has a certain fascination (you might like to light a candle and look

Figure 2.8
'I propose to bring you, in the course of these lectures, the Chemical History of a Candle .... There is no better, there is no more open door by which you can enter into the study of natural philosophy....'

* Faraday's lectures were originally published as M. Faraday, *A Course of Six Lectures on the Chemical History of a Candle*, Royal Institution of Great Britain, London, 1861.

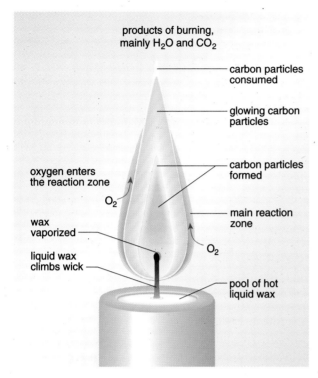

Figure 2.9
The structure of a candle flame.

Figure 2.10
A hydrocarbon molecule typical of that found in candle wax, $C_{30}H_{62}$. The molecule is represented in its fully extended form by a space-filling model.

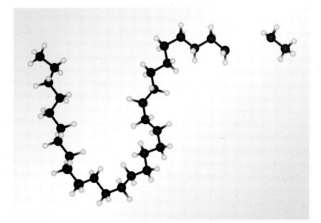

Figure 2.11
A ball-and-stick model of a wax-like hydrocarbon molecule being torn apart by breaking a carbon–carbon bond. The two fragments that result each contain a carbon atom which is involved in only three chemical bonds.

at the flame). In simplified terms, it can be described as a 'zone in which chemical reactions between gases are occurring'. These reactions produce both heat and light. The temperature of the flame varies from about 600 to 800 °C in the vicinity of the wick to temperatures in the region of 1 200 °C at its outer edges. A candle flame, in fact, tells us quite a lot about what we see when a whole variety of fuels burn. It is best described by dividing the 'volume' it occupies into regions, as shown in Figure 2.9.

The fuel of a candle is 'candle wax' which is largely composed of long-chain hydrocarbon molecules each containing mainly about 30 carbon atoms. The molecular formula for a straight-chain alkane containing 30 carbon atoms is $C_{30}H_{62}$ (Figure 2.10).

The application of a lighted match to the candle melts some of the wax and the liquid formed travels up the wick and is vaporized. (Once alight, the heat from the candle flame sustains this process.) But apart from melting and vaporizing the wax, the heat has a much more dramatic effect. Some of the hydrocarbon molecules in the vapour, and also in the surface layers of the molten wax, are literally torn apart, that is bonds between adjacent carbon atoms, as well as those between carbon and hydrogen atoms, are broken. The fragments that are formed, as illustrated in Figure 2.11, each contain a carbon atom that is not fully bonded. This means that they will be highly reactive species. Indeed, they will react fairly indiscriminately with any other fragment or molecule they come across, with the result that the number of highly reactive species increases. The overall result is 'chemical turmoil' and this spreads out, or diffuses, away from the wick of the candle.

Some of the hydrocarbon fragments that are formed become completely stripped of hydrogen atoms in the reactions that occur and give rise to species – again highly reactive – that contain only carbon atoms. These can combine together, usually towards the centre of the flame where the oxygen supply is limited, to give soot particles consisting of many thousands of carbon atoms. At the temperatures involved, these particles become incandescent and glow, so giving the flame its characteristic yellowish-white appearance.

The outer edges of the flame represent the so-called 'main reaction zone'. It is here that all of the various species, including soot particles, are eventually oxidized by oxygen in the surrounding air to give carbon dioxide and water vapour as the main products. The heat produced by the many different chemical reactions is responsible for the temperature of the flame. The blue to bluish-green tinge seen in the main reaction zone, but most visible close to the wick, is due to the presence of small molecular fragments which are highly reactive and short-lived: these emit coloured light when they are first formed.

> **Question 3**   In his lectures, Faraday inserted a glass tube into a candle flame and demonstrated that when a burning splint was brought up to the other end of this tube a flame appeared. This is shown in Figure 2.12. How would you account for this?

Figure 2.12
A simple experiment with a candle.

The candle flame is an example of a **diffusion flame**, that is one in which the rate of combustion is controlled by the rate at which the vaporized fuel and oxygen are transported to the reaction zone. Other types of flame result from the fuel and oxygen being premixed prior to combustion. Good examples are the flames from well-adjusted natural gas burners (Figure 2.3) or a butane gas burner which you may use for camping or plumbing. Soot formation in these flames is minimal because the ample supply of oxygen ensures complete conversion to carbon dioxide and water vapour.

To return to the candle flame, how do we write a chemical equation to describe the combustion process? This is not too difficult to achieve as long as we concentrate on the *overall* process. We know that the starting materials are candle wax and oxygen, and that the final products are carbon dioxide and water vapour. For simplicity, we can take solid candle wax to be composed mainly of the hydrocarbon $C_{30}H_{62}$. Given this simplification, the balanced chemical equation, although it looks complicated, can be written as

$$2C_{30}H_{62}(s) + 91O_2(g) = 60CO_2(g) + 62H_2O(g)$$

You can confirm that the equation does balance by noting that there are 60 carbon atoms, 124 hydrogen atoms and 182 oxygen atoms on both its left- and right-hand sides. Just thinking about the ingredients in the reaction suggests that the combustion could not proceed in a single step. Imagine what would have to happen! All of the chemical bonds in the fuel and oxygen molecules would have to break – at the same time – and then the individual atoms would have to recombine in such a way as to form just $CO_2(g)$ and $H_2O(g)$. This is so unlikely that it could never take place. Instead, and in a sense as we 'see' in the flame, the actual reaction proceeds in a number of discrete steps with each step involving a relatively minor rearrangement of chemical bonds. It is *all* of these individual steps taken together that accounts for the overall chemical transformation. The balanced chemical equation, *since it is only concerned with the starting materials and final products*, gives no hint of the underlying complexity.

The combustion of candle wax is a relatively complicated example, but it does illustrate a very general, and important, aspect of chemical reactions, combustion or not. For the majority of reactions, the overall chemical change, as described by a balanced chemical equation, is achieved by a number of intermediate steps. If we could view what was happening at the molecular level, then it would be these steps that we would see. The reactant species would be present, and products would eventually be formed, but in between there would be a number of quite distinct steps. The species (molecules, fragments of molecules, or ions) that are involved along the way are called **intermediates**. It is often the case that these intermediates are so reactive that their presence is exceptionally difficult to detect. The full set of all separate steps that taken together account for an overall chemical transformation from reactants to products is called the **reaction mechanism**. As already stated, the balanced chemical equation for a reaction gives no information about the mechanism by which it takes place. However, to understand a chemical reaction it is very important to try to find out as much as possible about its mechanism; often a difficult, but intriguing, task.

> **Question 4**   One of the preliminary discrete steps in the combustion of methane involves just the methane molecule. Try to suggest what this step will be.

Summary of Chapter 2

This Chapter has mainly concentrated on what a balanced chemical equation does *not* reveal about a chemical reaction. There are three important points: (a) chemical reactions either release heat to, or absorb heat from, their surroundings; (b) the speed of chemical reactions covers the whole range from being imperceptibly slow to explosively fast; and (c) the majority of chemical reactions occur by a series of discrete steps which involve intermediates. The intermediates involved in the mechanism of a chemical reaction can be very reactive and so their presence is often difficult to detect. The set of all the separate steps that lead to an overall chemical change is called the mechanism of the reaction.

Combustion is a reaction in which an element, or compound, burns in a plentiful supply of oxygen. In simplified terms, fireworks have two main ingredients. These are a fuel which burns to give an oxide and an oxidizer which provides the source of oxygen. The oxidizer has oxygen chemically bound within its structure.

The candle flame, which is an example of a diffusion flame, can be partitioned into distinct regions. These regions provide a good, general illustration of the manner in which flames are formed. In some types of burner, fuel and air are premixed prior to combustion.

Chapter 3
Energy: a closer look

3.1 A definition, and a problem of storage

As demonstrated in the previous Chapter, chemical reactions can either release heat to, or absorb heat from, their surroundings. A way of discovering more about this behaviour is to focus on chemical reactions and suggest that they involve energy changes. However, to take this idea further it is necessary to have a better appreciation of what we mean by energy. What is it? What are its properties?

A very familiar use of the term is in everyday conversation. For instance, how many times have you complained that 'I haven't got enough energy to do this'? Used in this way, we interpret energy as 'the something that enables something else to be done'. Moving to a more fundamental level, we are faced with the fact that energy is an abstract concept; it does not take up space, nor can it be put in a bottle for examination. You may also have at the back of your mind the fact that energy is related to mass; indeed, Einstein stated that energy is equal to mass multiplied by the velocity of light squared ($E = mc^2$). However, it is only in certain nuclear processes that the energy changes are sufficiently large so that the corresponding mass changes are measurable. We shall not consider such processes in this Chapter.

In practical terms, the concept of energy is of central importance in science because it provides a means of describing, and rationalizing, natural phenomena; for example, the melting of ice or the intricacies of a candle flame. A simple description of energy that acknowledges this practical impact is that 'Energy is the capacity to do something'. This is not too far-removed from our everyday interpretation of the term.

At first sight, this description, although it needs some refinement, may seem far more suited to mechanical, rather than chemical, situations. However, as you will see, closer examination shows that this is not the case. The word 'capacity' is important. It conveys the idea that energy is 'stored' and, by implication, it must be transferred in some way if it is 'to do something'. This fits in with everyday experience. For instance, the wound-up spring in a clockwork toy car stores energy. When the spring is released, it turns a series of cogs and the car moves forward and overcomes the frictional resistance of the surface on which it is resting. An alternative way of describing what is happening is to say that as the spring unwinds it is doing **work** and this results in the toy car moving forwards. In general, work is done whenever any object is moved or mechanically altered in some way.

So, these ideas allow us to say a little more about the 'doing something' in our description of energy. If work is involved, then it relates to 'moving or mechanically altering an object in some way'.

Energy has another important capacity. You might be able to suggest what this is if you think about dipping your toe in a hot bath. As you would be able to feel, your toe would become uncomfortably hot. Clearly, energy is transferred but in this case no work is done. In fact, energy has been transferred in a special, but not uncommon, way; it has been transferred as heat. The transfer arises because of the temperature difference between the bath water and your toe, i.e. heat flows from the hot water to your relatively cold toe. So, energy has the capacity to make things hotter as long as a temperature difference is involved.

Figure 3.1
An everyday energy store.

The ideas we have developed about energy are just as valid when expressed in chemical terms. Fuels, which after all are no more than chemical compounds, provide good examples. We pay 'energy bills' for the various fuels we use and essentially purchase them as stores of energy (Figure 3.1). When it is burned, a fuel can supply heat or, when it is burned explosively in a car engine, it can be used to do the work of moving a motor car forwards. The fuel itself is the store of energy *but the release of some of this energy is dependent upon a chemical reaction taking place.* In the same way, food is also a store of energy; but to release some of this energy it must be consumed and digested first.

But how is energy stored? In fact, there are only two main ways and you may be familiar with these already. If we consider a wound-up spring, then energy is stored because of the way the spring is compressed. Until the spring is released, there is no movement and we refer to the stored energy as **potential energy**. By contrast, a flowing river also stores energy, but in this case energy is associated with motion and it is referred to as **kinetic energy**. The release of some of this energy can cause damage, for example the river can erode its banks. The sum of all of the individual kinetic and potential energy contributions that characterize a particular situation is referred to, not surprisingly, as the **total energy**.

3.1.1 Kinetic energy

Kinetic energy, as already stated, is the energy that an object possesses by virtue of the fact that it is in motion. Kinetic energy depends upon *both* the speed and the mass of the object; e.g. a fast-moving and swollen river does more damage than the trickle of a sedate brook. In chemistry, the gaseous state provides an important example of the application of ideas associated with kinetic energy. As you know, in the gaseous state the molecules are moving about in a chaotic fashion and so on an individual basis each molecule will have a kinetic energy associated with it. If the molecule has a mass m, and is travelling at speed v, its kinetic energy can be calculated from the formula $\frac{1}{2} \times m \times v^2$. The higher the speed at which the molecule travels, then the larger is its kinetic energy, or, put another way, it 'stores' more energy. It's worthwhile looking at this in a little more detail.

The movement of a molecule in a gas is dominated by the collisions it experiences with other molecules. At one instant, a collision may increase its speed while, at another, an unfavourable collision can bring it to a virtual standstill. This means, as shown schematically in Figure 3.2, that at any instant in a gas the molecules will all be moving around at different speeds and so each molecule will have a different kinetic energy. The number of collisions that a molecule experiences is immense. For example, the oxygen molecules in the volume of air you have just breathed in will *each* be involved in about a billion collisions every second. With such a large number of collisions taking place, the *overall* spread of molecular speeds in a gas will always remain the same, although of course individual molecules will behave in a highly unpredictable manner.

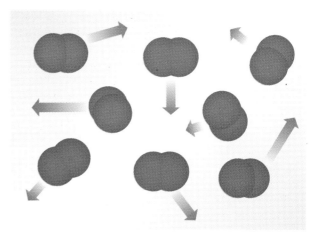

Figure 3.2
Molecules in a gas, e.g. oxygen, moving about randomly and at different speeds.

The form of the spread of molecular speeds for any gas at a particular temperature was first worked out by the Scottish scientist James Clerk Maxwell towards the end of the 19th century. He produced a formula which allows the percentage of the total molecules present, which have speeds in a given range, to be calculated: this is now known as the **Maxwell distribution**. It is conveniently demonstrated using a histogram. The calculated distribution for oxygen gas at room temperature is shown in this way in Figure 3.3. On the horizontal axis, the molecular speeds are broken down into small increments of 20 m s^{-1}. Each interval has a vertical height, marked by a horizontal bar. The value that this horizontal bar represents gives the percentage of all of the molecules present whose speed falls within the interval in question.

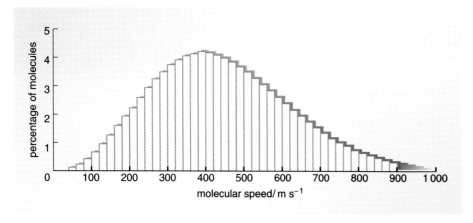

Figure 3.3
The Maxwell distribution for oxygen gas at room temperature plotted in the form of a histogram. The molecular speeds on the horizontal axis are divided into increments of 20 m s^{-1}.

Question 5   Which interval of molecular speed has the largest percentage of oxygen molecules associated with it in the histogram in Figure 3.3? If we could somehow measure the speeds of individual oxygen molecules in a sample of the gas at room temperature, what can you say about the probability of finding oxygen molecules with speeds within this particular interval?

At any instant in a sample of oxygen gas, the individual molecules have distinct kinetic energies. These kinetic energies will be tiny, however, because the molecular mass is so small, i.e. the term m in the equation $\frac{1}{2} \times m \times v^2$ is very small. It is far more convenient for comparative purposes to consider a sample containing one mole of oxygen molecules.

■ How many molecules of oxygen are there in a sample containing one mole of O_2 molecules?

■ The number will correspond to the value of the Avogadro constant, i.e. there will be 6.02×10^{23} O_2 molecules.

The *total* kinetic energy in a gaseous sample containing one mole of O_2 molecules will be just the sum of all of the kinetic energies of the individual oxygen molecules; this could be derived, although we shall not do so, from the Maxwell distribution. This total amount of kinetic energy is quite modest. For instance, it is equivalent to the amount of energy needed to raise the temperature of one litre of water by about one degree Celsius. Alternatively, it is a small amount when compared to the energy typically released in a combustion reaction – for instance, it corresponds to roughly the energy produced when a few drops (no more than 0.1 g) of petrol is burned.

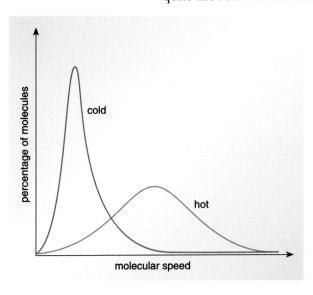

Figure 3.4 shows in schematic form the Maxwell distribution for a sample of a gas at two different temperatures. For convenience, rather than plot the individual histograms, smooth curves have been used to represent each distribution. As you can see, all speeds occur at both temperatures, but at the higher temperature there is a relative decrease in the percentage of slow-moving molecules and a relative increase in the percentage of rapidly moving ones.

Figure 3.4
The Maxwell distribution for a gas at two different temperatures.

■ What do you think would happen to the total kinetic energy of a gas sample if the temperature was increased?

■ Intuitively, and as discussed in Book 1, you would expect that the molecules, on average, would move around more quickly; that is they would become more 'thermally agitated'. This is confirmed by the distribution shown in Figure 3.4. Increasing the temperature, therefore, will increase the total kinetic energy of the sample.

If the temperature of a gas is raised, then it will store more kinetic energy; by the same token, on cooling down the gas will release this energy.

3.1.2 Potential energy

Potential energy is a more difficult type of energy store to appreciate. It is the energy that any object possesses because of the position in which it finds itself. It is sometimes, descriptively, referred to as 'hidden energy' because no movement is involved. The ornaments that are on shelves in your home store this type of energy because they have the capacity to do something: to fall and break, to damage the floor, to damage your foot. The further an ornament has to fall, then the greater its store of potential energy. It is important to recognize that it is the interaction between the ornament and the Earth, that is gravity, that 'gives' the ornament its potential energy. This idea of an 'interaction' provides the key to understanding potential energy. It is because an object experiences interactions – and gravity is a familiar one to us – that it stores potential energy. In fact, an ornament no matter where it is placed will always experience gravity and so will always possess potential energy due to this type of interaction: the exact amount will depend upon where it is placed.

Chemical compounds also act as stores of potential energy. This energy arises because the atoms, molecules or ions from which they are constructed interact with one another – through the various types of forces that were described earlier in Book 1. We can illustrate this idea using a relatively simple compound. Figure 3.5 shows the model structure that was used in Book 1 to demonstrate how the sodium and chloride ions are arranged in the structure of sodium chloride.

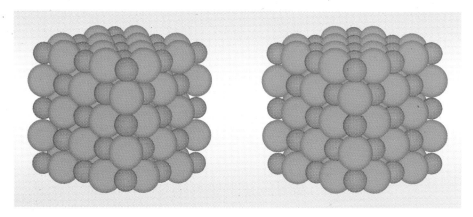

Figure 3.5
A model of the structure of sodium chloride taken from Book 1.

▨   How would you describe the main forces in the sodium chloride structure?

■   The main forces are ionic. A given Na^+ ion is attracted to oppositely charged Cl^- ions, and repelled by other Na^+ ions. Similarly, a given Cl^- ion is attracted to oppositely charged Na^+ ions, and repelled by other Cl^- ions.

It is because the ions in the structure interact with one another that each ion has an 'interaction' or potential energy associated with it. Taken as a whole, the crystalline ionic compound is a store of potential energy because of these ionic interactions.

▨   Look carefully at the structure of sodium chloride in Figure 3.5 and focus on just one horizontal plane of ions. What happens if this plane is moved in any horizontal direction you may choose?

■   You should have been able to see that no matter what direction was selected, the movement brings identically charged ions closer to one another. This increases the repulsive ionic interactions at the expense of the attractive ones.

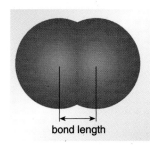

Figure 3.6
In an oxygen molecule, the bond length is typically 120 picometres (pm).

Another way of expressing the answer to the question is to say that the effect of moving the plane horizontally is to *raise* the potential energy. Overall, in sodium chloride, the structure is such that a balance is achieved among the various ionic interactions. In fact, sodium chloride adopts a structure in which the potential energy is at a *minimum.*

Covalent bonding also acts as a store of potential energy. In the case of an oxygen molecule, there is a particular value of potential energy which is associated with the two oxygen atoms when they are at a distance apart equal to the normal bond length as shown in Figure 3.6.

If the oxygen atoms are 'pushed' closer together, it turns out that the potential energy increases and, similarly, if they are drawn apart the potential energy also increases: the behaviour is not unlike that of a simple coil-spring. It can be represented by plotting a **molecular potential-energy curve** as illustrated in Figure 3.7. In this Figure, the horizontal axis represents the separation between the two oxygen atoms; it represents a distance. The vertical axis represents potential energy.

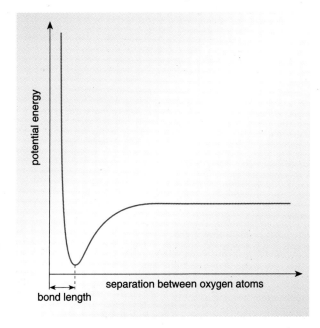

Figure 3.7
A schematic representation of the molecular potential-energy curve for the oxygen molecule. The minimum in the potential energy curve corresponds to the bond length of the oxygen molecule.

    What situation is represented by the curve at the far right of the plot?

    At the far right of the plot, the two oxygen atoms are separated by a large distance and consequently the interactions between the two atoms are extremely weak. However, the potential energy is high; remember that this type of energy depends upon position.

The situation when the two oxygen atoms are far apart can be taken as the reference point for the molecular potential energy curve. As the two oxygen atoms approach one another, the potential energy falls. Eventually, a chemical bond is formed and it is in this situation that the potential energy has a minimum value. At smaller separations, the nuclei of the two oxygen atoms get very close and very strong repulsive interactions come into play; the potential energy rises sharply.

3.1.3 Total energy

Although it is not immediately obvious, you should now begin to realize that any material, whether solid, liquid or gas, acts as a store of energy by virtue of its microscopic make-up. The total energy that is stored is simply equal to the sum of the various kinetic and potential energy contributions. Since the energy is stored within the sample, it can alternatively be referred to as **internal energy**. In a sample of methane gas, for example, there will be contributions to the internal energy from the kinetic energy due to the motions of the individual molecules as well as from the potential energy associated with the chemical bonding.

In principle, the internal energy of a sample contains a whole variety of contributions – right down to those involving the kinetic energies of the electrons and those associated with the way in which the atomic nucleus is constructed. However, it is *changes* in internal energy that are important because, in practice, it is these that we monitor. Indeed, it is not too large a step to suggest that when reactants are converted into products in a chemical reaction, it is changes in internal energy that have occurred. Since these changes are chemical in origin, we shall often refer to them as changes in **chemical energy**. This is a subject we shall have a lot more to say about in due course.

3.2 Conservation and transformation

In the previous section, we used ornaments on shelves in your home to introduce the idea of potential energy. You may, at some time, have been unfortunate enough to knock an ornament off a shelf. How would you describe what happened in energetic terms?

As it left the shelf the ornament would have picked up speed and so therefore it increased its kinetic energy. At the same time, as it fell, its potential energy would be decreasing. This strongly suggests that there is a relationship between the increase in kinetic energy and the decrease in potential energy of the ornament. But what is this relationship?

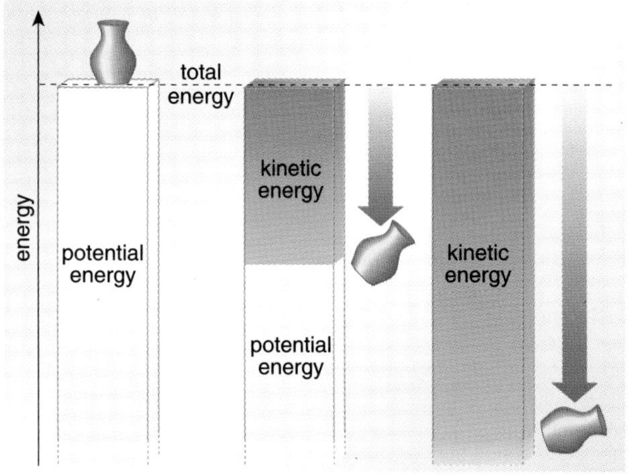

We can answer the question, although it relates to only a minor event, in terms of a property that is of universal importance. At every stage during the fall, the sum of the kinetic energy and potential energy will have remained constant, in other words the total energy also remained constant. This is shown schematically in Figure 3.8.

The idea that total energy does not change is embodied in a law of central importance in science, namely the **law of conservation of energy**. It can be expressed as:

> *'Energy can neither be created, nor destroyed '.*

A simpler form, using a minimum of words, is that *'energy is conserved'*. Taken to its extreme, the law implies that the total energy in the Universe

Figure 3.8
As an ornament falls, there is an increase in kinetic energy and decrease in potential energy, but the total energy remains constant. On impact with the floor, the ornament (or its fragments) will still possess potential energy due to gravitational attraction. For simplicity, the schematic view focuses only on the ornament as it falls.

Box 3.1 Computers and energy calculations

The power of modern computers, coupled with an increased understanding of how to calculate realistically the interactions between atoms, molecules and ions, has meant that computer modelling is becoming a widely used technique in various areas of molecular and materials science. One modelling approach is to calculate the 'internal potential energy' for a trial structure (e.g. that of a biological molecule, synthetic polymer or inorganic material), and then search for a structural arrangement that minimizes the potential energy as this would be the most likely structure. Figure 3.9 shows one example of this approach.

This Figure is concerned with a class of materials called zeolites. These are crystalline compounds which are aluminosilicates, that is their main structural framework is constructed from the elements aluminium, silicon and oxygen. The arrangement of the elements is special, in that the aluminium and silicon atoms are always

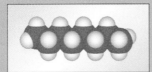

(a)

(b)

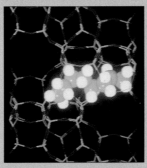

(c)

Figure 3.9
Modern computer modelling. (The difference between (a) and (c) in colour coding of the octane molecule is not significant.)

present in 'rings' held together by oxygen atoms. A typical arrangement is shown in Figure 3.10.

About 36 distinct framework structures, some beautifully intricate,

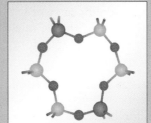

Figure 3.10
Two aluminium (darker grey) and four silicon atoms held together in a ring by oxygen atoms. As indicated, each of the aluminium and silicon atoms will also take part in other rings. Zeolites containing various ring sizes are known.

are known for zeolites. An important property of all of the structures is that, on the molecular scale, they have an internal open porous structure with large cavities linked together, or connected by a network of channels. This is illustrated in Figure 3.9b which shows a computer-generated view of the internal structure of a complex, but commercially important, zeolite called 'ZSM-5'. The porous nature is clearly seen. (This Figure also demonstrates another important role of computers – the ability to focus on, and illustrate, a key feature of a complex

structure.)

Since the mid-1960s, zeolites have been extensively employed in commercial chemical processes, particularly in the petrochemical industry. This is because they act as catalysts, that is they speed up otherwise slow chemical reactions while remaining unchanged themselves. But more than this, the reactions they speed up occur *within* the zeolite and so the porous structure controls – essentially on the basis of size – the nature of the products that are formed. To understand these processes it is essential to gain information on how molecules fit into zeolite crystals. Computer modelling is one way of doing this. For example, Figure 3.9c shows how, according to a 'minimum-energy calculation', the hydrocarbon molecule octane (Figure 3.9a) 'fits' into the ZSM-5 framework. Such calculations, and the information they provide, are being used in the development of improved zeolites for more efficient industrial processes.

is constant. Or, in other words, the Universe has a fixed amount of energy which always remain the same – before, during and after *any* type of change that may occur. On a more mundane level, it tells us that the heat produced when we reacted aluminium metal with bromine liquid did not represent the creation of energy, nor did the fall in temperature when potassium nitrate was added to water represent the destruction of energy. *We can be certain that chemical reactions cannot create or destroy energy.*

To return to ornaments for a final time, it is, of course, reasonable to ask how a particular piece gained its potential energy. The obvious answer is that it was put there! A more scientific answer is that 'muscular energy' was used to place it on the shelf; in other words, muscular energy was transferred to the ornament and so was transformed into potential energy. Can you anticipate the next question?

Where did the muscular energy come from? In biochemical terms the answer is complex but essentially it relates back to food. The processes which break down food in the body are chemical in nature and so involve changes in chemical energy. Some of this chemical energy is transformed into mechanical energy when we use our muscles.

You should now be beginning to appreciate that when we ask the question 'Where did a particular type of energy come from?' the answer inevitably involves a chain of energy events. In many cases this chain will lead back to the energy that has been received on Earth from the Sun's radiation. This is discussed further in Chapter 9.

So, in summary, although energy can neither be created nor destroyed, it can be transformed from one form into another. In general, in most processes, energy is transferred or transformed through a whole series of events and different forms of energy are involved.

Activity 1 A chain of events

Figure 3.11 shows, in schematic form, the chain of events involved in bringing an electric kettle full of water to the boil. As an exercise, briefly consider the chain of energy transformations that are involved. You may have to do a little research, e.g. by using an encyclopedia, to find out what happens in a coal-burning power-station.
Figure 3.11
Boiling an electric kettle.

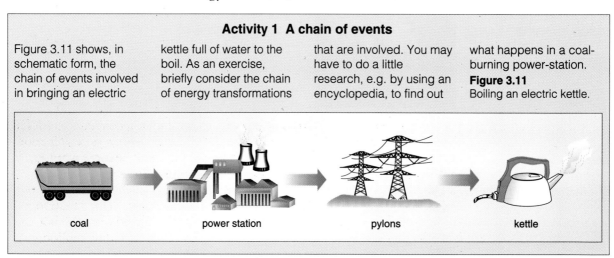

coal   power station   pylons   kettle

3.3 Some like it hot: temperature and heat

So far, we have used the term 'heat' in a fairly informal way; that is, we have used it as we would in everyday conversation. However, at this stage it is useful to find out more about its underlying meaning.

To begin with, a distinction should be made between temperature and heat. Temperature is simply a measure of 'hotness', it can be gauged by touch or more practically by taking temperature readings using a thermometer of some kind. If it turns out, for example, that the temperature of a tumbler of water is 15 °C then this should not be confused in any way with the amount of energy that would be required to raise the temperature to, say, 20 °C. This would depend only upon the quantity of water in the tumbler and would be the same whether we were raising the temperature from 15 °C to 20 °C, from 60 °C to 65 °C, or any other 5 degrees Celsius range on the temperature scale (as long as the water did not boil or freeze). Similarly, if we wanted to lower the temperature by 5 °C, we would have to extract the same amount of energy as that found necessary to achieve the corresponding temperature rise.

Until well into the 19th century, heat – or as it was then known *caloric* – was thought to be a 'fluid' that flowed from one body to another. We now know that this view was incorrect, but the picture portrayed by 'heat flow' still remains extremely useful. To say exactly what **heat** is, however, is not straightforward. For our purposes, and as we touched on earlier, we shall think of it as a transfer of energy that results from temperature differences. This view is consistent with a molecular picture as illustrated in Figure 3.12.

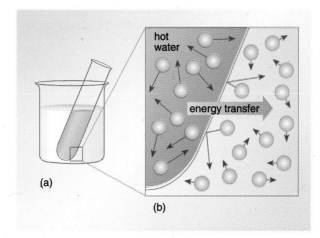

Figure 3.12
(a) A test tube of hot water in a beaker of colder water and (b) a schematic view (at the molecular level) of the situation in (a).

Figure 3.12a shows a test tube of hot water standing in a beaker of colder water. Common experience tells us that the water in the test tube will cool down and that in the beaker, at least to begin with, will become warmer; in other words we say that a transfer of heat occurs. But a more correct statement is to say that the *internal energy* of the water in the test tube decreases, while that of the water in the beaker increases. This situation is shown schematically in Figure 3.12b. Energy is transferred from the hot water to the cold water. As a consequence, the random chaotic motion – or thermal motion – of the water molecules in the test tube decreases whilst that in the surrounding colder water increases. In other words, the internal energy lost by the hotter water is the same as that gained by the colder water.

So, for instance, when we talk about 'heating water' in an electric kettle, what do we really mean? The phrase is really shorthand for the more cumbersome statement 'transferring energy from a hot electric element to the colder water because of the temperature difference, thereby increasing the water's internal energy and so causing its temperature to rise'. As you will no doubt agree, it is far more convenient to continue with everyday informality. Thus, it is perfectly acceptable to talk about 'the heat produced by a chemical reaction' as long as it is recognized that a transfer of energy is involved. However, informality can be taken too far. It would not be correct to refer to the heat in the water in a kettle, because what would really be meant was the internal energy of this water.

Box 3.2 William Thomson, Lord Kelvin (1824–1907)

William Thomson was born in Belfast and at the age of eleven entered Glasgow University where his father held the chair of mathematics. He subsequently went to Cambridge University, and then spent a brief time in Paris, before returning to Glasgow (1846) to occupy the chair of Natural Philosophy: he was then 22 years old. He was primarily a mathematician and physicist, but with some very practical interests. Thus, in the 1860s, his work on the electrical properties of cables contributed significantly to the success (in 1866) of the first transatlantic communication cable. He was duly knighted in the same year and later (1892) created a peer with the title Lord Kelvin. He was very much concerned with the intellectual struggles that led in the 1850s to the recognition of the central importance of energy and its conservation.

Figure 3.13
Lord Kelvin presenting his last lecture at Glasgow University, 1899.

Question 6   Overall, this Chapter has taken a closer look at energy. Without looking back, and before reading the Summary, see if you can answer the following three questions.

(a) How would you describe energy?

(b) What are the two main forms of stored energy?

(c) What are two important properties of energy?

Summary of Chapter 3

In everyday terms, energy is 'something' that enables 'something' else to be done. In fundamental terms, it is an abstract concept, but more practically it is the capacity to do something. This can be moving or mechanically altering an object in some way; alternatively, if a temperature difference is involved it can be making an object hotter.

A fuel when it is burned can supply heat, or be used to do work as in a car engine. The fuel itself is the store of energy but the release of some of this energy is dependent upon a chemical reaction taking place. Kinetic energy is one way of storing energy; it is the energy that an object possesses because of its motion, and can be calculated from the formula $\frac{1}{2} \times m \times v^2$. Potential energy is the other way of storing energy; it is the energy that an object possesses because of its position.

The spread of molecular speeds at any instant in a gas is described by the Maxwell distribution: this distribution depends upon temperature. In between collisions in the gas phase (at a particular temperature), individual molecules will have distinct kinetic energies. The total kinetic energy in a gas sample is the sum of all of the individual kinetic energies; it increases with increasing temperature.

Chemical compounds store potential energy. In a simple ionic compound such as sodium chloride, there is a potential energy contribution due to interactions between the ions. A molecular potential-energy curve can be drawn for a molecule such as oxygen; the minimum in this curve represents chemical bond formation. Total energy is the sum of kinetic and potential energy. For a chemical compound, it is referred to as the internal energy. Computer modelling, for instance to find structural arrangements that minimize the potential energy, is becoming a powerful technique in chemical studies.

The law of conservation of energy states that energy can neither be created nor destroyed, that is 'energy is conserved'. Thus, chemical reactions can neither create nor destroy energy. Energy can be transformed from one form into another. In general, in most processes, energy is transferred or transformed through a whole chain of energy events. Heat is the transfer of energy that results from temperature differences. The description that 'heat is produced by a chemical reaction' should be taken to imply that a transfer of energy is involved.

Question 7 'Which is the more important to society: a supply of food or a supply of energy?'. What points would you make, given that it is expected that the answer is going to be open-ended?

Chapter 4
Practical matters

In the preceding Chapters, we have not considered energy in any quantitative way. Here, therefore, we redress this situation and turn to the question of how to describe quantities of energy. In fact, we shall answer this question in two distinct ways. To begin, we shall be specific and consider the units in which energy is measured. A much wider perspective will then be taken as we look at the quantities of energy that are available to us from fossil fuels. As the Chapter develops, we shall also take a more careful look at the way in which measurements are reported and, just as importantly, how they are used in calculations.

4.1 Units of energy

As you should have found in the Activity, there are different ways in which energy can be measured. In practical, and usually familiar, situations, this may be acceptable, but in scientific terms it can cause confusion. The ideal is to select a single preferred unit which is consistent with the recommendations of the SI system.

For energy, the preferred unit is the **joule**; it is given the symbol J. The name honours the English scientist James Prescott Joule (1818–1889), the son of a Manchester brewer and a contemporary of Lord Kelvin. Joule's scientific work contributed a great deal to our modern understanding of heat and energy.

The joule may be the preferred unit, but what do we know about it? For example, 'How large is one joule?'. A favourite textbook example is that it is roughly equivalent to the kinetic energy that the famous apple had when it hit Newton's head; or, in the same vein, it amounts to the energy you would use to pick a fallen apple from the ground. Alternatively, each beat of your heart uses about one joule of energy as it drives blood round your body. In relative terms, the joule represents a fairly small quantity of energy.

A popular energy unit that is still widely used (but is not in the SI system) is the **calorie**, abbreviation cal. It derives from the Latin word for 'heat' and has the attraction of a very practical definition: 1 cal is the amount of energy that is required to increase the temperature of 1 g of water by 1 °C. It is worth noting that another unit called the Calorie, which starts with a capital letter, is often used when discussing dietary requirements: this is equal to 1 000 calories, or 1 kilocalorie.

The fact that the joule represents a fairly small quantity of energy often means that energy requirements, or energy changes, turn out in practice to be large numbers when expressed in this unit. For this reason, it is often the case that a 'powers-of-ten' notation is used or, even more conveniently, a prefix is used that denotes a power of ten. Table 4.1 provides a summary, although you are asked in Question 8 to complete some of the details for yourself.

**Activity 2
How is energy measured?**

Quite a lot of information can be discovered about the ways in which energy is measured by looking around your own home; for example, look at gas bills, electricity bills, the meters themselves, food labels and any other source you think may be relevant. What do you find?

Table 4.1 Different quantities of energy.

energy	1 000 J	1 000 000 J	1 000 000 000 J
powers-of-ten notation			
prefix notation	1 kilojoule, in symbols 1kJ	1 megajoule, in symbols 1MJ	1 gigajoule, in symbols 1GJ

Question 8 In Table 4.1, the entries for the 'powers-of-ten notation' are omitted. As revision of the material you studied in Book 1, complete the Table.

▪ The unit of the therm, on which gas bills were previously assessed, is roughly equivalent to 105 000 000 J. How would you write this in prefix notation?

▪ Converting to a powers-of-ten notation gives 105×10^6 J, so that the therm is 105 MJ. (This is much easier to write.)

Since the calorie is still widely used, it is useful to know how it relates to the joule. Some idea of the relationship can be gained by looking at food labels. For example, on a pack of drinking chocolate (Figure 4.1) the energy content per 100 g serving is stated as both 367 kcal and 1 554 kJ. The joule is thus the *smaller* unit and 1 cal is approximately equal to 4 J.

In fact, there is an *exact* relationship between the calorie and the joule and so a definition can be made. It is*

$$1 \, cal = 4.184 \, J$$

Figure 4.1
A typical food label showing energy content.

* Similarly, this is the case for certain other SI and non-SI units; thus, the inch is defined exactly in terms of the centimetre by the relationship, 1 in = 2.54 cm.

Question 9 When you added potassium nitrate to a test tube of water in Experiment 2.15, you felt the solution become colder. In fact, in a similar experiment, it was found that adding close to 1.20 g of potassium nitrate to a volume of water of 100 cm³ caused the temperature to fall by about one degree Celsius. Approximately how much energy (expressed in J) would be required to raise the temperature of the final solution in this experiment so that it was the same as the starting temperature of the water at the outset? (For simplicity, take the mass of the solution to be 100 g and assume that the solution requires the same quantity of energy as pure water to raise its temperature by a given amount.)

Figure 4.2
Domestic electricity consumption is measured in kWh.

As you will see in Chapter 5, the calculation you have just carried out in Question 9 has a direct bearing on how we formally describe the energy change associated with dissolving potassium nitrate in water.

Before leaving this section, there is one final unit to consider. This is the **kilowatt-hour**, which as you have seen in Activity 2 has the abbreviation kWh. It is the unit in which we purchase electricity for our homes (Figure 4.2), but it is also used as a general unit for comparisons between different sources of energy on a national scale. It is not immediately obvious that this 'compounded' unit represents a certain number of joules.

The problem is that it is based upon the definition of another unit: the **watt**, symbol W. This unit describes the *rate* at which energy is transformed from one form into another, or transferred from one store to another. For example, in your own home, if you monitor your electricity meter as you turn on more electrical appliances you will find that it indicates that more electricity is being used in a given time period; that is, energy is being used at a faster rate. The rate at which energy is being used is referred to as power and is measured in watts. The advice to 'turn off the power' before doing any electrical repair is not only sound, but has some scientific basis.

One watt represents the transformation or transfer of 1 J over a time period of 1 s, i.e. 1 W is 'one joule per second' or 1 J s^{-1}. It represents a relatively small rate of energy change. So, how do we interpret the unit of a kilowatt-hour?

One approach is to re-express it in words; it represents a 'rate of energy provision of 1 000 joules per second which lasts for a time period of one hour'.

▓ Now try to calculate how much energy expressed in J is represented by 1 kWh.

▓ In one hour, there are 3 600 s and in each of these seconds 1 000 J is provided. The total number of joules represented by 1 kWh is therefore $1\,000 \times 3\,600 \text{ J} = 3\,600\,000 \text{ J}$, or in a more compact notation 3.6 MJ.

Activity 3 Determining the relationship between joules and kilowatt-hours

If you have a particular type of electricity meter in your home, it is possible to use it to determine the relationship between kilowatt-hours and joules for yourself. The meter needs to be of a type which allows small fractions (hundredths) of kWh consumed to be measured. The old-fashioned (non-digital) type shown in Figure 4.3 is ideal.

In outline, the idea behind the experiment is to measure the fraction (in hundredths) of a kWh that is needed to bring an electric kettle filled with cold water to the boil. How would you

design the experiment and what would you need to measure? Can you suggest any areas of uncertainty that might affect your measurements? (Remember, other devices in your home may be using electricity at the same time as your experiment.)

Figure 4.3
A 'suitable' electricity meter. The rotating disc, which is part of the mechanism, makes '150 revolutions per kWh used' according to the label on this meter. The outer rim of the disc is conveniently marked so that the number of revolutions can be counted.

If you have a suitable electricity meter, you may wish to carry out the experiment discussed in Activity 3. Even if you cannot do this, you should still spend a little time working through the analysis of the experimental results.

> **Experiment 2.16 The relationship between joules and kilowatt-hours: boiling an electric kettle**

4.2 Handling measurements

All scientific measurements, no matter how carefully they are carried out, give results which have uncertainties associated with them. Thus, repeating any experiment several times, even though the starting conditions are exactly the same, will be expected to give a spread of results rather than a single unique value. Broadly speaking, uncertainties arise because of the way an experiment is designed, the fact that human judgement may be involved and, inevitably, that measuring instruments have their limitations. Some of these factors were discussed in Activity 3. Depending on circumstances, the uncertainties in an experiment may be large or small and an important question that arises is, 'How do we deal with them?'.

4.2.1 Precision and accuracy

Figure 4.4 shows two cartoons of an archer practising. In the first cartoon, the archer has fired a number of practice arrows at the bull's eye and achieved reasonable success.

Figure 4.4
Archery practice on
(a) a still day and (b) a day
on which the wind is steady
and strong.

■ Can you suggest two distinctive features of the practice session that account for its 'reasonable success'.

■ First, the arrows fired are clustered closely together. Secondly, if we take an 'average position' for all of the shots, and take it as the result, then it is in the ring defining the bull's eye – the intended target.

In fact, we can say the archer has been both precise and accurate. Precise, because all of the arrows fired are clustered closely together, and accurate because the average of the shots is close to the actual target.

The notions of **precision** and **accuracy** can also be applied to scientific measurements. When a measurement is repeated a number of times, then the distribution of results that arises is put down to **random error**. It is this type of error which gives rise to the variation from measurement to measurement, with sometimes a result being high and other times low. When the random error is small, we say the measurements are precise. So, precision is determined by the uncertainty associated with random error. It is also a quality that can be directly assessed as it is determined by the spread of results.

What determines accuracy? The second cartoon, although it stretches the analogy a little, provides an answer. In this case, the archer is practising in a steady, strong wind. The archer remains precise – because the arrows are all clustered together – but the average position for all of the shots is now very much displaced from the intended target. The archer is inaccurate. The accuracy has been affected by the continuous presence of the strong wind; this has caused a **systematic error**. In the same way, the accuracy of scientific measurements can also be affected by systematic errors. For example, dust on a balance pan would introduce a systematic error into the measurement of the mass of an object. Accurate measurements are those which have small systematic error and give results which are close to the true value. So, accuracy is determined by the uncertainty associated with systematic error.

■   Could you look at a set of measurements and assess the accuracy?

■   To do this it would be necessary to have a good idea of the true value; without this knowledge, the accuracy could not be assessed.

In practice, the accuracy of a measurement depends very much on the quality of the apparatus and the design of the experiment. The archer can easily adjust to new conditions, but it is often very difficult to be sure of, and eliminate, the sources of systematic error when making a scientific measurement.

> **Question 10** Using the archery analogy, sketch, and briefly justify in words, the distribution of arrows for imprecise and inaccurate practice shots at the bull's eye.

4.2.2 A matter of significance

The Egyptian tour guide told his visitors that the pyramid they were on their way to visit was 4 507 years old. 'Four thousand five hundred and seven!' remarked someone. 'Yes', said the guide. 'When I first started here I was told it was built 4 500 years ago, and I've been here for seven years now.'

A quantity of energy equal to 59.5 J was found to raise the temperature of the polyethylene sample by 6.28 °C. Thus, the amount of energy required to raise the temperature of the sample by one degree Celsius is:

$$\frac{59.5\,\text{J}}{6.28\,°\text{C}} = 9.474\,52\,\text{J}\,°\text{C}^{-1}$$

Figure 4.5
4 507 years old!

These two examples, one anecdotal, and the other science-based, illustrate that care must be taken with the way in which we interpret numerical information. In the first case, common sense should tell us that the statement '4 500 years ago' is not accurate and should not be taken too literally. In the second, the calculator has provided a number of decimal places for the answer, but can we arbitrarily choose how many to quote? The purpose of this section is to outline, but only briefly, 'good housekeeping rules' for reporting and using scientific measurements.

The simple task of reading the temperature from a thermometer (Figure 4.6) provides a good example.

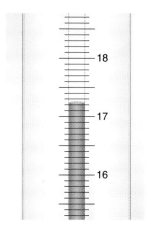

Figure 4.6
Measuring a temperature.

- What is the temperature according to the thermometer in Figure 4.6?

- Your answer will undoubtedly have been 17.2 '*and a bit*' °C but you were probably unsure as to what to do about the '*bit*'. If you did do something, your reading most likely was in the range 17.22 to 17.25 °C.

It is not too difficult to imagine that there would be a spread of results if a number of people were asked, independently, to take the temperature reading.

- Does this spread of results reflect the precision or the accuracy of the readings?

- It reflects the precision which is associated with the uncertainty due to random error.

To be strict, the accuracy of the measurement should also be considered. For example, there could be poor thermal contact between the thermometer and the material whose temperature was being monitored and so taking a reading without giving sufficient time for the temperature to settle could introduce a systematic error. For any scientific measurement, it is important to be aware of the uncertainties, but for 'everyday discussion' such detail becomes cumbersome. For this reason, a simple convention, involving the idea of **significant figures**, is introduced.

The convention is that the value of a measurement is quoted in such a way that *only* those digits which have *real significance* are included. This means that all digits, *up to and including the first uncertain digit*, are listed: these are called the significant figures in the measurement. Thus, if a temperature of 17.25 °C is reported, we recognize that this measurement has *four* significant figures. We can be certain of the '17.2 *part*' and we know that the uncertainty lies in the last digit, but even so this last digit is 'a good guess'. Experimental results should always be quoted with the appropriate number of significant figures consistent with what is known about the uncertainties in the measurements.

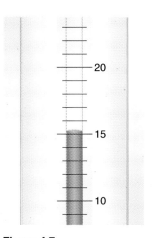

Figure 4.7
Another temperature measurement.

- Figure 4.7 shows a temperature measurement using an inexpensive thermometer. What temperature would you report?

- The temperature is 15. '*something*' °C and so should be reported to three significant figures. A reasonable measurement would be 15.2 °C or 15.3 °C.

Box 4.1 Significant figures: some common problems

One common problem that arises when using significant figures centres on the interpretation of the digit 0 in a number. This can be illustrated by considering the four reported energy measurements opposite.

The powers-of-ten notation is very useful for indicating (or deciding upon) significant figures. For instance, writing 0.082 J as 8.2×10^{-2} J shows clearly that the measurement is to two significant figures. It also confirms, as stated opposite, that 'initial zeros do not count'. As another example, re-expressing 900 J as 9.0×10^2 J would indicate

79.0 J	There are three significant figures. The zero has the same significance that any other digit would have in this position.
50.6 J	There are three significant figures. Again the zero counts as an ordinary digit.
900 J	It is not clear how many significant figures are intended here! If it were three, then the energy measurement would be, say, around 899 to 901 J. Alternatively, the measurement could simply represent the nearest ten, or even hundred, of joules. The interpretation really depends on the context in which the number is being used.
0.082 J	There are two significant figures because *initial zeros do not count in this context*.

clearly that the measurement was to two significant figures. *If in doubt about significant figures, always try the powers-of-ten notation.*

It should be noted that it is the number in a quoted value of, say, 32.6 kJ that indicates the measurement is to three significant figures. To convert the measurement into joules, it should be written in a powers-of-ten notation, that is 3.26×10^4 J, so that the three significant figures remain obvious (32 600 J would be ambiguous!).

Finally, it has to be remembered that some numbers have to be treated as *exact*. If you like, they have an infinite number of significant figures. Thus, in the definition 1 cal = 4.184 J the number 4.184 is in this category; it could be written as 4.184 00 … with the zeros continuing for as long as you wish to write them.

Question 11 How many significant figures are in the measurements (a) 10.1 °C; (b) 600.5 J; (c) 62.0 kJ; (d) 0.15 °C?

Of course, the results of measurements are often used in calculations. In this case, two basic 'rules-of thumb' should be taken into account. Both rules act only as rough guides and are based upon the idea that it will be the measurement with the greatest spread of results (the least precise measurement) that will determine the overall precision of the final answer. The rules are:

Addition and subtraction. When numbers are added or subtracted, the number of decimal places in the result should be the same as in the measurement with the *least* number of decimal places. This rule assumes that the numbers are not expressed in scientific notation.

Multiplication and division. When numbers are multiplied or divided, the number of significant figures in the result should be the same as in the measurement with the *fewest* significant figures.

To use either rule, it is necessary to know how to 'round' a number. This is a procedure for rewriting a number so that it is correct to a given number of decimal places, or to the nearest whole number, e.g. ten, hundred, etc. Suppose, for example, we are dealing with the number 52.673. To round to the nearest whole number, the procedure is to look to the right of the position in question. If the digits (irrespective of decimal point) lie in the range 0000 … to 4999…, then the number is 'rounded down'; if they lie in the range 5000… to 9999…, then the number is 'rounded up'. So, 52.673 to the nearest whole number is 53 because 673 lies in the latter range and so '52' is rounded up to '53'. Rounding is always carried out in a single step. The procedure is quite general: 52.673 to two decimal places is 52.67, to one decimal place it is 52.7, and to the nearest ten it is 50.

■   Two pieces of aluminium are weighed on different balances. The results are 7.34 g and 3.6 g, respectively. What is the total mass?

■   The numerical sum is 10.94 but according to the rule-of-thumb this should be rounded to one decimal place. The total mass should be reported as 10.9 g.

   Question 12   At the start of this section, a calculation concerning a sample of polyethylene was quoted. How would you report the final answer?

Special care has to be taken when considering multiplications (or divisions) which involve integers; that is, exact whole numbers. For example, one way of calculating the total mass of two objects each weighing 2.17 g would be to find 2×2.17 g. In fact, it is far better to think of this as an addition and so use the rule-of-thumb for addition; that is 2.17 g + 2.17 g = 4.34 g. As you should be able to see, this way of thinking ensures that there are always two decimal places in the quoted final mass – no matter how many 2.17 g objects are involved.

■   If there were 22 objects each weighing 2.17 g, what would be the total mass?

■   47.74 g. (If we had thought in terms of a multiplication, then we might have been tempted to round to three significant figures and so quote the answer, inappropriately, as 47.7 g.)

The ideas underlying significant figures are useful since they allow 'good housekeeping practice'. However, it is often convenient – say in demonstrating how to do a calculation – to use less than the 'usual' number of significant figures for the measured quantities involved. In fact, this is an approach we have already used in this and the previous Book. For instance, only three significant figures have been used for relative atomic masses (in particular, that for hydrogen has been taken as 1.01 rather than 1.0080). In general, we shall continue with this approach but, from now on, we shall be more explicit about the way in which we carry out the 'rules for calculation'.

4.3 Fossil fuels: reserves, resources and lifetimes

The fossil fuels – coal, oil and natural gas – are sometimes referred to as primary energy sources and are responsible for most of the energy provision on which modern society depends. It is salutary to remember, however, that these fuels are **non-renewable energy resources**: once used, they are gone forever. In discussions concerning the quantities of fossil fuels recoverable from the Earth, the terms **resources** and **reserves** are often used. The meaning of these terms must be treated with caution.

Resource measures the *total* quantity of the particular fossil fuel *in* the Earth; to make this clearer it is sometimes referred to as 'amounts in place'. It is not a fixed quantity: on the one hand, it is reduced by consumption; but, on the other hand, exploration can reveal entirely new resources of fossil fuel.

Reserves, or sometimes as they are more informatively called 'proved reserves', are generally taken to be 'those quantities which geological and engineering information indicate with reasonable certainty can be recovered in the future from known sources under existing economic and operating conditions'. The difference between the estimates of resources and reserves can be large. For instance, the World Energy Conference in 1989 put the amount in place of coal in the United Kingdom as 378 billion tonnes but the recoverable reserves as only about 4 billion tonnes. These figures differ also, depending on the sources of the information.

As is to be expected, estimates of reserves change depending on the state of exploration and extraction technology. In addition, the economic and political circumstances of the day as well as the prices of competing fuels play a role. Political factors are highlighted in Figure 4.8 which shows a map of the world indicating the proved reserves of oil at the end of 1993 for the major oil-producing areas. The overwhelming dominance of the Middle East, coupled with possible political instability in this part of the world, serves to emphasize the caution to be exercised when discussing 'world' oil reserves.

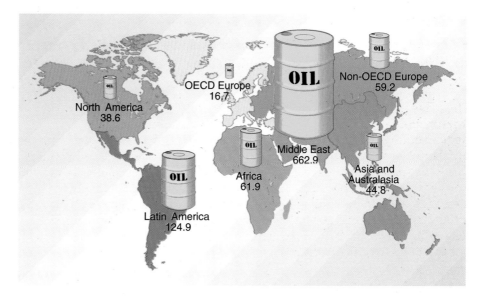

Figure 4.8
The proved reserves of oil at the end of 1993. The reserves are measured in 'thousand million barrels'; the term 'barrel' is defined in Figure 4.9. The Middle East holds more than 65% of the total proved reserves.

In national statistics, large quantities of energy are often expressed in 'tonne equivalents' or 'million tonne equivalents' of coal or oil. The oil industry invariably favours a barrel as a practical measure of petroleum products. A few facts and figures are summarized in Figure 4.9.

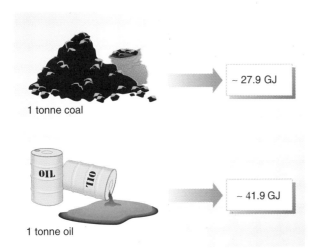

1 tonne coal → ~ 27.9 GJ

1 tonne oil → ~ 41.9 GJ

Figure 4.9
A tonne of coal equivalent (tce), if based on 'average coal', is deemed to have an 'energy content' of the order of 27.9 GJ; this corresponds to the energy that would be produced when 1 tonne of this coal is burned in air. In the same way, a tonne of oil equivalent (toe), again based on an 'average' conversion factor, has an energy content of the order of 41.9 GJ. Roughly speaking, 1 tonne of oil can supply the same amount of energy as 1.5 tonnes of coal. The volume of a barrel of crude oil is roughly 160 litres (or in old-fashioned units 35 imperial gallons). The number of barrels per tonne depends on the density of the product and this varies quite markedly from heavy crude to light petroleum spirit. A commonly accepted average figure is about 7.3 barrels per tonne. The photograph shows oil being transported in barrels by the Oil Creek Valley Railroad in Rouseville, Pennsylvania, in the early years of the industry.

Question 13 What would you estimate to be the energy (expressed in J) stored in the proved reserves of oil in the Middle East according to the 1993 statistics (Figure 4.8)?

Box 4.2 A few further energy statistics

There are lots of different energy statistics. Just a few are given opposite (but it is not necessary to remember them).

- In national energy statistics, natural gas is quoted by the cubic metre (m³) and this volume of gas has an energy content of approximately 38 MJ.
- Power stations express their output in megawatts (MW) or, for very large stations, gigawatts (GW).
- A litre of petrol can provide about 10 kWh which is equivalent to the energy required to run a single bar electric fire for 10 hours.
- A tonne of coal can provide about 7 700 kWh.
- Annual world energy consumption (strictly, transformation into other forms!) in the early 1990s was roughly 8 billion tonnes of oil equivalent.

4.3.1 Two important everyday fuels: natural gas and coal

Figure 4.10
The photograph shows an early evening view of the Arabian Desert and Gulf area photographed by the crew of the Space Shuttle *Columbia* during a mission in 1990. The white areas are cities and the orange lights seen throughout the region are gas flares from oil exploration and production facilities both on- and offshore. In 1978, the total amount of gas flared worldwide was estimated to be equal to 130 million tonnes of oil equivalent. In recent years, the waste of such a precious resource has been reduced considerably.

Natural gas and coal are two fuels which are used, essentially, 'straight from the ground' without any major chemical refinement. Table 4.2 summarizes the *proved* reserves of these fuels, on a worldwide basis, at the end of 1993.

Table 4.2 Proved reserves of natural gas and coal at the end of 1993.

Area of the world	Natural gas[a] (trillion cubic metres)	Coal (billion tonnes)
OECD Europe[b]	5.3	96.9
USA	4.7	240.6
Canada	2.7	8.6
Latin America	7.6	11.4
Africa	9.7	62.0
Middle East	44.9	0.2
Non-OECD Europe[c]	57.1	315.5
Asia and Australasia	10.0	303.9
TOTAL WORLD	142.0	1039.1

[a] 1 000 cubic metres of natural gas is roughly equivalent to 1.5 tonnes of coal equivalent. Thus, the *column of figures* for natural gas proved reserves, multiplied by 1.5, can also be taken to represent billion tonnes of coal equivalent.

[b] OECD Europe: Austria, Belgium, Denmark, Finland, France, Germany, Greece, Iceland, Italy, Netherlands, Norway, Portugal, Republic of Ireland, Spain, Sweden, Switzerland, Turkey, United Kingdom.

[c] Non-OECD Europe: Albania, Bulgaria, Czech Republic and Slovakia, Hungary, Poland, Romania, former Soviet Union (incl. Azerbaijan, Kazakhstan, Russian Federation, Turkmenistan, Ukraine and Uzbekistan), Yugoslavia and former Yugoslav Republics, Cyprus, Gibraltar and Malta.

Natural gas

Natural gas is a component of '**petroleum**' (literally meaning 'rock oil').
This latter term is generally employed to describe the wide range of
hydrocarbons that are extracted from the ground as gas and crude oil
deposits. It is almost universally accepted that natural gas (as well as crude
oil with which it is often associated) derives from the remains of plant and
animal life that lived in the oceans millions of years ago. The composition
is variable, but as shown in Table 4.3 the most important component is
methane gas. One of the most important non-hydrocarbons is hydrogen
sulfide; molecular formula H_2S. This gas smells of 'rotten eggs' and is
highly toxic. A natural gas contaminated with H_2S causes pollution
problems and is termed a 'sour' gas; a 'sweet' gas has less than one part
per million of hydrogen sulfide.

Table 4.3 Some constituents of natural gas.

Substance	% of total volume
methane	85–96.0
ethane	2–6
propane	0.5–5
other hydrocarbons	<4 total
carbon dioxide	0–5
helium gas	0–0.5
hydrogen sulfide	0–5

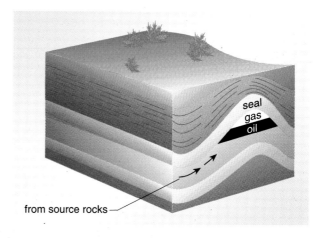

Natural gas can occur almost entirely on its own (in the technical language
of the oil industry, it is 'non-associated') or be associated with crude oil,
either dissolved within the oil or in a layer above it (Figure 4.11). It is
estimated that about 72% of the world's proved gas reserves are non-
associated. As Table 4.2 shows, the Middle East has a sizeable store of the
world's natural gas proved reserves but this is still significantly less than its
store of the proved oil reserves as illustrated in Figure 4.8.

Estimates of the world's recoverable natural gas reserves are subject to
uncertainties and, hence, so is the remaining 'life' of this resource. Proved
reserves have actually increased since the 1970s as a result of exploration
and changing economic factors; one source suggests an increase of three
times the 1970 figure. UK production of natural gas in 1993 was about 63
billion cubic metres and consumption was about 67 billion cubic metres.
The estimated (1993) proved UK reserve is 600 billion cubic metres and so
on these figures the reserve will be exhausted in a matter of years, unless
supplemented from elsewhere. Worldwide, despite all of the uncertainties,
the length of time that the remaining reserves of natural gas will last if
production continues at the 1993 rate is almost certainly measured in tens
rather than hundreds of years. For Western Europe, the average is about 25
years, although for Norway by itself the estimate is over 70 years. It is
worth noting that some geologists believe that there are truly massive
amounts of methane buried deep below the Earth's surface.

Figure 4.11
A common type of
petroleum reservoir in which
the oil and gas have
separated into different
layers. Petroleum is formed
in deep source rocks and
then migrates over time, by
processes which are still not
fully understood, into
reservoir rocks such as
porous sandstones.
Eventually, the migration
process leads to the
petroleum accumulating in
geological traps. It is worth
noting that, as one author
puts it, 'Extracting oil is
more like squeezing treacle
out of a brick than lifting
bucketfuls of water from a
well'.

Coal

As described earlier in Part 1 of this Book, coal is formed from the remains of plant material. The formation process involves a number of stages which finally result in anthracite being formed if the conditions are right. In chemical terms, the process of coal formation results in a progressive increase in the relative carbon content of a coal; for example, high quality anthracite contains over 90% by mass of carbon although the chemical structure still remains complex.

Figure 4.12 compares the length of time that the remaining reserves of oil, natural gas and coal are expected to last in major areas of the world if production continues at 1993 levels. The overall world situation is also summarized in the Figure: global coal reserves will considerably outlast the combined reserves of oil and natural gas. Coal certainly has a lifetime measured in several hundreds of years but it has to be said that the mix of economic, political and environmental issues which currently surround it make any future predictions difficult, particularly for a given country. In the UK for example, coal production was at a 225 million tonne peak in 1953 but in 1993 it was less than one-third of this amount; in contrast, in other parts of the world production continues to rise steadily. Environmental opposition to coal as a fuel is growing, but the arguments are complex and summaries therefore can be misleading. It is the case, however, that coal combustion on a worldwide basis does release more carbon dioxide – a so-called greenhouse gas (see Chapter 9) – into the atmosphere than the combustion of natural gas, and secondly it does cause air pollution, particularly because of its sulfur content.

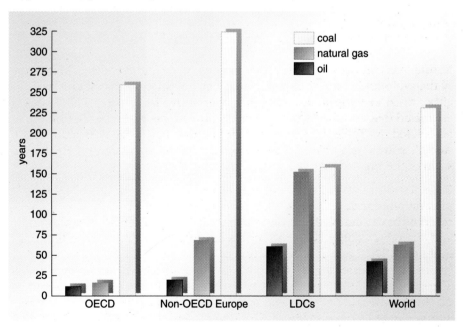

Figure 4.12
A matter of time! A comparison of the length of time that the remaining reserves of different fossil fuels are expected to last if production continues at 1993 levels. Countries in OECD Europe and Non-OECD Europe are listed beneath Table 4.2. Latin America, Africa, Middle East and Non-OECD Asia are included as LDCs.

About half of the sulfur in British coals is in the form of pyrite (FeS_2) and this is partly removed during the cleaning operations in coal preparation plants prior to combustion. Most of the non-pyritic sulfur is chemically bound in the complex molecules that form coal (Figure 4.13). Combustion of coal results in virtually all of the remaining sulfur being converted to sulfur dioxide gas, SO_2. This gas is harmful to health because it irritates the lungs. However, the most far-reaching problem of SO_2 in the atmosphere (but by no means all of it from coal combustion) is due to its high solubility in rainwater to form a mixture of sulfur-containing acids, e.g. sulfuric acid. These increase the *natural* acidity of rainwater – there can be an increase of up to '100 times' in hydrogen ion, H^+, concentration – and contribute towards 'acid rain'. It is important to note that not all of this effect is due to SO_2 emissions because nitrogen oxides (some from coal combustion) also play a role due to the corresponding formation of nitrogen-containing acids, e.g. nitric acid. A large proportion of the SO_2 produced rises into the atmosphere and is naturally dispersed by prevailing winds to create acid rain days later, and many hundreds of miles away. For example, there is strong evidence that SO_2 emissions from the industrial regions of Britain and Central Europe are responsible for ecological damage in Scandinavian countries. Strategies for reducing SO_2 emissions during coal combustion can be very successful (Figure 4.14).

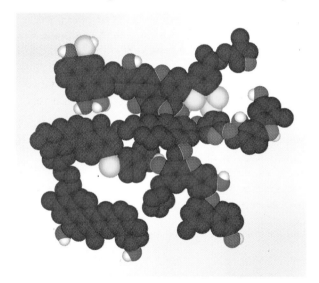

Figure 4.13
A space-filling representation of part of the molecular structure for coal. The structure is very complex, as well as containing the elements sulfur (S) and nitrogen (N).

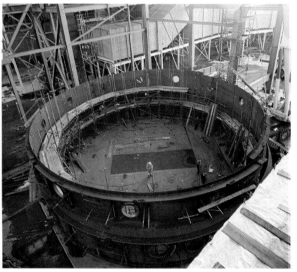

Figure 4.14
So-called flue gas desulfurization (FGD) is one method of reducing SO_2 emissions from coal combustion. In this process, flue gases are reacted with a limestone (calcium carbonate) slurry in a large 'scrubber' unit. The result is that the 'cleaned flue gas' has only about 10% of its original SO_2 content. The gypsum (hydrated calcium sulfate) slurry that is also produced has a number of industrial uses, including wallboard manufacture, or it can be used, with some restrictions, as a landfill material. In economic terms, the installation of FGD technology adds about 10% to the capital and operating costs of a power station. The photograph shows the construction of a 'scrubber' unit at the Drax power station in the UK.

Summary of Chapter 4

The preferred (SI) unit of energy is the joule, symbol J. An often-used but non-SI unit is the calorie, abbreviation cal. One calorie is the amount of energy that is required to increase the temperature of 1 g of water by 1 °C. There is an exact relationship between the calorie and the joule: 1 cal = 4.184 J. The joule represents a fairly small amount of energy and so, in practice, energy changes are often expressed using a powers-of-ten or prefix notation; common units are kJ, MJ and GJ. The watt describes the rate at which energy is being used; it represents the transformation or transfer of 1 J over a time periods of 1 s. The unit 1 kWh (i.e. 1 kilowatt for 1 hour) is equivalent to 3.6 MJ.

Precision and accuracy in experimental measurements are determined by the uncertainties associated with random and systematic errors, respectively. Precise measurements are those for which the random error is small. Precision can be directly assessed because it relates to the spread of results obtained by a number of repeated measurements. Accurate measurements are those for which the systematic error is small and the results are thus close to the true value. In practice, the accuracy of a measurement depends critically on the quality of the apparatus and the experimental design.

Only those digits that have real significance should be included when reporting the value of a measurement. The number of significant figures is the number of digits up to, and including, the first uncertain digit. A common problem when using significant figures centres around the interpretation of the digit 0 in a number. Writing a number in powers-of-ten notation often helps in deciding on the number of significant figures.

Two basic 'rules-of-thumb' should be adhered to when using the results of measurements (quoted to a number of significant figures) in calculations. In addition and subtraction, the number of decimal places in the result should correspond to the measurement with the least number of decimal places. In multiplication and division, the number of significant figures in the result should be the same as in the measurement with the fewest significant figures. Special care has to be taken when carrying out multiplications (or divisions) which involve integers.

Resource measures the total quantity of a particular fossil fuel in the Earth and reserve measures the quantity of the fuel which, with reasonable certainty, could be recovered in the future from known sources. Neither is a fixed quantity. Future exploration could reveal new resources. New extraction technology or changing economic and political conditions could change the estimate of reserves.

National energy statistics use 'energy measures' including 'tonne of coal, or oil, equivalents'. Barrels are the measurement used for petroleum products. The main constituent of natural gas is methane; coal has a complex molecular structure. Worldwide, the 'projected life' of natural gas is almost certainly measured in tens of years, while coal has a lifetime measured in several hundreds of years.

Activity 4 Global energy calculations

It is generally accepted that the figure for the flow of energy from the Sun intercepted by the Earth's surface is of the order of 178 million gigawatts (1.78×10^{17} W). Figure 4.15 shows how the average amount of sunlight falling on a horizontal area of the Earth's surface over a 24-hour period varies with location.

(a) How does the energy reaching the Earth from the Sun compare to the annual world energy consumption in the early 1990s (roughly 8 billion tonnes of oil equivalent)?

(b) Your calculations should suggest that the direct use of solar energy is a very attractive proposition. What factors do you think must be taken into account when considering how to utilize this source of energy? (Most current research focuses on the transformation of solar to electrical energy.)

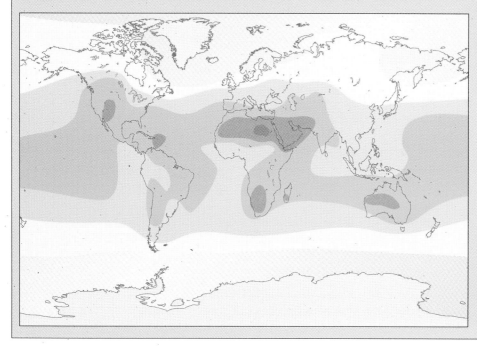

▨	300 W m^{-2}
▨	250 W m^{-2}
▨	200 W m^{-2}
□	150 W m^{-2}
□	100 W m^{-2}
∿	below 100

Figure 4.15
Sunlight at the Earth's surface.

Chapter 5
Chemical processes and energy changes

The idea that chemical reactions involve energy changes has already been established. However, very little has been said about how these changes can be treated in a quantitative way. At the moment, for instance, we are not in a position to compare the merits of different fuels on an 'energy-produced' basis. The purpose of this Chapter is to provide the background to change this situation. As a first step, it is useful to introduce a simple classification.

5.1 A means of classification

Figure 5.1
(a) At the start, the temperature of the water was close to 24 °C.
(b) Dissolving potassium nitrate in water; for this experiment, the *fall* in temperature can be seen to be about 5 °C.
(c) Dissolving lithium chloride in water; for this experiment, the *rise* in temperature can be seen to be about 7 °C.

When potassium nitrate was dissolved in water, you found that the temperature of the solution fell. This was simply interpreted as indicating that the dissolving process takes in heat from the local surroundings, including the solution itself. If lithium chloride (LiCl) had been used instead, then you would have obtained the opposite result: there would have been a rise in the temperature of the solution because it turns out that the dissolution of LiCl in water produces heat. The results of typical experiments are shown schematically in Figure 5.1.

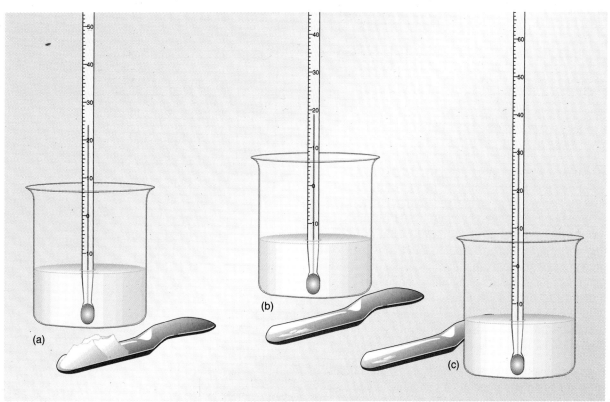

The two experiments, although relatively simple, introduce the idea that it should be possible to classify all chemical processes in terms of their energy changes, that is they either absorb heat from, or release heat to, their surroundings. However, before taking this step it is important to be clear how the term '**surroundings**' should be interpreted. This may mean that you will have to change slightly your own view of this term.

The term, of course, implies that there is something to surround – this is always called the **system**. It is the system which is the part of the world in which we have a specific interest – it could be a solution containing a simple ionic compound dissolved in water, or the reaction mixture containing aluminium metal and bromine liquid. *Everything* else is the surroundings, but to be more practical it could simply be a beaker and the air surrounding it. The two terms are illustrated schematically in Figure 5.2.

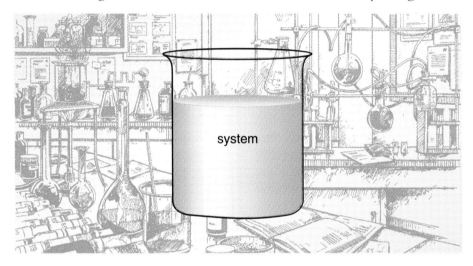

Figure 5.2
A schematic view of a system and its surroundings. Note carefully that the beaker is part of the surroundings.

So, now we can make the classification:

● A process that releases heat into the surroundings is called **exothermic**.

● A process that absorbs heat from the surroundings is called **endothermic**.

The terms exothermic and endothermic are from the Greek: *thermo* meaning heat, *exo* meaning outside and *endo* meaning within. Hence *exo*thermic conveys the idea that heat is given *out* and *endo*thermic that heat is taken in. Clearly, all combustion reactions – or any other reactions that burst into flames – are exothermic. In more general terms, however, the actual temperature rise produced by an exothermic reaction depends on a number of factors – for instance, the conditions under which it takes place and how fast the reaction proceeds. The reaction between iron and moist air to produce hydrated iron oxide, or more familiarly the brown flakey substance we call rust (Figure 5.3), is in fact a very exothermic reaction. Ordinarily, this rusting process takes place so slowly that the liberation of heat is not detectable.

For an endothermic reaction, it is worth repeating that heat is absorbed from the surroundings. If, for example, you hold a container in which such a reaction is occurring it will feel cold because, in effect, your hand becomes part of the surroundings and heat is absorbed from it.

Figure 5.3
Rusting iron, an exothermic but very slow reaction.

5.2 Enthalpy

The idea that chemical processes are accompanied by changes in internal energy is an important one. Thus, if we take the dissolution of solid lithium chloride in water as an example, we could write a 'thermochemical equation' in the following way:

$$LiCl(s) = Li^+(aq) + Cl^-(aq) + energy$$

The equation shows that the reaction is exothermic. Energy (in the form of heat) is released to the surroundings. This is because the solution containing lithium and chloride ions has a lower internal energy than that of the starting materials – solid lithium chloride and water. In these terms, it is possible to represent what is happening in a pictorial fashion as shown in Figure 5.4. In this picture, the vertical axis represents energy.

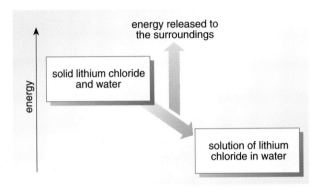

Figure 5.4
A pictorial view of the energy changes that occur when solid lithium chloride dissolves in water.

In practice, as we know, when lithium chloride dissolves the energy released will raise both the temperature of the solution that is formed *and* the surroundings. Of course, in time, the solution will cool down to the starting temperature and so all of the energy released will eventually dissipate into the surroundings. In this sense, there is an ambiguity in understanding what the 'energy-released arrow' means in Figure 5.4. Does it refer to the situation immediately after lithium chloride dissolves, or sometime later? But, if the latter, how much later? The ambiguity can be removed by thinking about the process in a different way. Picture in your mind's eye adding solid lithium chloride to water to give a final solution that had exactly the same temperature as the starting materials – the process is shown schematically in Figure 5.5. It is a *constant temperature* process. In this case, you can be certain there is no ambiguity – *all* of the energy produced is transferred to the surroundings.

Figure 5.5
Dissolving LiCl in water at constant temperature.

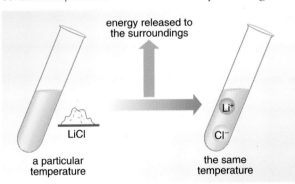

So, thinking about a chemical process as though it occurs at constant temperature is a very useful device because it removes any ambiguity in interpreting what is meant by the energy change.

■ What happens when an endothermic reaction occurs at constant temperature?

■ In this case, we can be certain that *all* of the energy absorbed will be transferred *from* the surroundings.

More often than not, chemical reactions, including those involved in natural processes, occur in a way that is open to the atmosphere; that is, in effect, they occur at *constant pressure*. For some chemical reactions this means that we have to be more careful about the way in which we interpret the change in internal energy. In Book 1, for example, you saw that ethyne (acetylene), C_2H_2, can be made by pouring water onto calcium carbide, CaC_2. This can be expressed chemically by the equation,

$$CaC_2(s) + 2H_2O(l) = C_2H_2(g) + Ca(OH)_2(s)$$

The reaction is exothermic but since ethyne is a gas the reaction will also be accompanied by a net increase in volume. Such a volume change has to be paid for. In other words, part of the internal energy change for the reaction is expended in pushing back the atmosphere in order to 'make space' for the ethyne gas. Thus, if we view the reaction as though it occurs at constant temperature, then the energy released as heat is not a direct measure of the internal energy change accompanying the reaction.

In most practical situations, however, it is the 'heat' that is released (or taken in) by a chemical reaction that is of interest, irrespective of any volume changes that may occur. In order to be specific, a new quantity is introduced: it is called enthalpy. (It comes from the Greek words for 'heat inside'.) The **enthalpy change** for any reaction at constant temperature – *as long as it also occurs at constant pressure* – can be taken to be a direct measure of heat changes. We simply ask you to accept that the stipulation of constant pressure ensures that enthalpy and heat changes are synonymous.

The symbol that is given to enthalpy is H. In practice, it is changes in enthalpy that are of interest and these are written as ΔH where the Greek letter Δ (spoken 'delta') is used to mean 'change of'. Values of ΔH can be measured experimentally and, indeed, it is on this basis that we shall compare different fuels in the next Chapter. (Activity 6 at the end of this Chapter gives you the chance to look at a simple experiment for measuring enthalpy changes.)

If we return to dissolving lithium chloride in water, we could imagine describing this process in words. We would have to say that the reaction is exothermic and that the enthalpy change, ΔH, corresponds to a certain number of joules. Furthermore, we would have to state the quantities of material involved because, as we shall see later, this affects the magnitude of the enthalpy change. This is all fine, but it does become tedious. One way to avoid having to state whether a reaction is exothermic or endothermic every time is to introduce a convention. We can design it so that it relates to the sign that we attach to the number of joules that ΔH represents. The sign is determined by focusing on what happens to the system. So,

●   For an exothermic reaction, the value of ΔH is taken to be negative (ΔH is less than zero). This is because the system releases heat to the surroundings; in effect, the system is *minus* some energy.

●   For an endothermic reaction, the value of ΔH is taken to be positive (ΔH is greater than zero). This is because the system gains heat from its surroundings; in effect, the system is *plus* some energy.

▓   If it was found that the heat released when 1.00 g of methane gas was burned in oxygen was 50.3 kJ, what would be the enthalpy change for this particular reaction?

▓   The reaction is exothermic and so, by convention, $\Delta H = -50.3$ kJ when 1.00 g of methane burns in oxygen.

▪ If you look back to Question 9 (p.200), what would be the approximate enthalpy change when 1.20 g of potassium nitrate was dissolved in 100 cm³ of water?

▪ Dissolving potassium nitrate in water is obviously an endothermic reaction. In Question 9, you estimated the energy required to raise the temperature of the solution so that it was the *same* as the starting temperature of the water in the experiment. If we assume that this energy is *all* that is required *from the surroundings*, then this energy will also be the same as we picture in our mind's eye to ensure that the reaction occurs at constant temperature. Thus, approximately, $\Delta H = +400\,J$ when 1.20 g of potassium nitrate dissolves in 100 cm³ of water under the conditions used in the experiment.

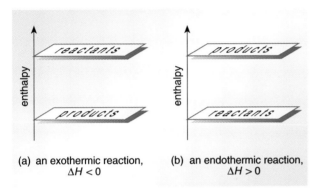

(a) an exothermic reaction, $\Delta H < 0$

(b) an endothermic reaction, $\Delta H > 0$

Figure 5.6
Schematic 'enthalpy ladders' for (a) an exothermic and (b) an endothermic reaction.

It is often useful to represent enthalpy changes for a reaction in a simple diagrammatic manner: an 'enthalpy ladder' can be used for this purpose. This is shown in Figure 5.6. Basically, a vertical 'enthalpy axis' is drawn and the combined enthalpies of reactants and of products are represented by 'rungs' at different levels. For reactions which involve only solids and liquids, the internal energy changes and enthalpy changes are always very similar to one another and so the vertical axis could just as well be labelled 'energy'. It is only in the case of reactions which involve gases in which there can be significant volume changes that the differences become apparent. We shall put enthalpy ladders to good use in Chapter 7.

Activity 5 A simple experiment for measuring an enthalpy change

Various simple home-based experiments have been described to measure enthalpy changes for reactions. One of these is illustrated in Figure 5.7.

The purpose of this experiment is to measure the enthalpy change when a given amount of methylated spirits is fully burned in air to give carbon dioxide and water.

The results of a typical experiment were:

● volume of water in heat-resistant glass beaker = 250 cm³;

● observed rise in temperature of the water in the beaker = 9.0 °C;

● mass of methylated spirits burned = 0.52 g.

Use this information to calculate the value of ΔH when the given amount of methylated spirits is burned in air. Where do you think the major uncertainties in the experiment occur? You should note that inspection of the underside of the beaker at the end of the experiment showed that there were some deposits of soot.

Figure 5.7
A simple home-based experiment to measure the enthalpy change for a reaction.

Summary of Chapter 5

This short Chapter has considered in a more formal manner the energy changes associated with chemical processes. Chemical reactions can be classified as exothermic or endothermic. The former release heat into the surroundings while the latter absorb heat from the surroundings. These classifications depend on a clear understanding of the terms system and surroundings. The system is the part of the world in which we have a direct interest; the surroundings is everything else.

It is useful to think about chemical reactions as though they occur at constant temperature; there is then no ambiguity as to what is meant by the magnitude of an energy change. The enthalpy change (ΔH) of reaction occurring at constant pressure and temperature provides a direct measure of heat changes. A useful convention is that for an exothermic reaction the numerical value of ΔH is represented as a negative quantity; for an endothermic reaction, it is represented as a positive quantity. The value of the enthalpy change for a given reaction depends on the quantities of material used. An 'enthalpy ladder' is a useful diagrammatic means of representing the enthalpy changes for a chemical reaction.

Chapter 6
Enthalpy changes in practice

Enthalpy changes can be measured experimentally for a whole variety of processes. In this Chapter, we focus on just two types of change, both of which are of considerable practical importance. To begin with, we look at enthalpy changes that accompany changes of state and so take the description started in Book 1 a little further. As you might guess, the main example will be that ubiquitous substance: water. Picturing events at the molecular level will prove to be very valuable in accounting for the enthalpy changes that occur. The focus of attention will then turn to chemical fuels. What is a good fuel? How can we establish a 'good fuel guide'? These questions raise a number of issues of which a key one relates to 'enthalpy changes'. The other issues that arise are fairly broad-ranging – you might like to pre-empt matters a little and think now what they might be. For instance, 'Are coal, oil and natural gas the only fuels worth considering?'; 'Why do we need highly refined petrol for our cars?'; 'How do living species store their energy?'; 'Are storage and transport of a fuel important considerations?'.

6.1 Changes of state: ice, water and steam

Boiling a kettle of water (Figure 6.1) is a very familiar process, indeed, it is so familiar that we usually don't bother to state that it's water we are boiling. Usually once a kettle has boiled (i.e. when the water within it has reached its boiling temperature: 100 °C at normal atmospheric pressure), we turn it off; but if we forget to do this, we know that it will 'continue to boil' and eventually 'boil dry'.

Figure 6.1
For a kettle containing one litre of water, it takes roughly seven times more energy to boil it dry than it does to boil it in the first place: it is not sensible to leave a kettle boiling.

■ If you were to measure the temperature at any stage during the continuous boiling process, what would you find?

■ The temperature would remain at 100 °C. In other words, despite the fact that energy is being supplied, the temperature doesn't rise.

What is happening, as described in Book 1, is that at a temperature of 100 °C – and as we continue to supply energy – the water is undergoing a change of state; we call it vaporization. The change of state is from a liquid to a gas, although the latter is often referred to as steam in the case of water.

■ If we represent the change of state by the equation
$H_2O(l) = H_2O(g)$
does this represent an endothermic or an exothermic process?

■ For vaporization to occur at the constant temperature of 100 °C, energy has to be supplied. Hence, the process is endothermic.

The enthalpy change associated with vaporization is simply called the **enthalpy of vaporization**. We can represent it as ΔH (vaporization), or just ΔH (vap) for short.

Why should vaporization be an endothermic process? Can you think of a way of putting forward an explanation? Since energy has to be supplied, it must be the case that the water molecules are 'more tightly held' in the liquid, as opposed to the gaseous, state. But why?

The reason is to be found at the molecular level. As you know, strong attractive forces due to hydrogen bonding play a major role in determining the properties of liquid water. In the liquid, individual molecules will always be moving about and so the hydrogen bonds they are involved in will forever be breaking and reforming. But on a larger scale this turmoil will simply resolve into an ever-changing network of hydrogen bonds. It is the cohesion of this network that has to be overcome if a water molecule is to escape into the gaseous state. That hydrogen-bonding is so important is emphasized by the fact that liquid water has the largest enthalpy of vaporization, *measured per gram*, of any known liquid.

If we return to the equation that represents the change of state for the vaporization of water, i.e.

$$H_2O(l) = H_2O(g),$$

then it could be interpreted as representing the situation when a *single* water molecule in the liquid 'breaks free' and enters the gas phase. Equally, the equation would still be valid no matter how many water molecules were considered. In particular, if we choose Avogradro's number, then the equation can be interpreted as

one mole of H_2O molecules in the liquid state

$$\xrightarrow{\text{is converted to}}$$

one mole of H_2O molecules in the gaseous state.

This is equivalent, as you have seen in Part 1, to giving the equation a *molar interpretation*. This idea is very useful as it also allows us to devise a means of showing what the enthalpy change is when one mole of H_2O molecules in the liquid state is completely vaporized. Experimentally (at 100 °C and 1 atmosphere pressure), it is found that ΔH (vap) = +40.7 kJ for this quantity of water. So, we can now write a **thermochemical equation**; it is

$$H_2O(l) = H_2O(g) \qquad \Delta H \text{(vap)} = +40.7 \text{ kJ}$$

This thermochemical equation conveniently describes the enthalpy change when one mole of H_2O molecules in the liquid state is fully vaporized. Alternatively, it would be just as valid to state the enthalpy of vaporization for water as ΔH (vap) = +40.7 kJ mol^{-1}.

> **Question 14** Table 6.1 compares the normal boiling temperatures and enthalpies of vaporization of liquid water and liquid methane. Briefly, how would you account for the differences?

Table 6.1 Liquid water and liquid methane.

Substance	Normal boiling temperature	Thermochemical equation	
H_2O	100 °C	$H_2O(l) = H_2O(g)$	$\Delta H(\text{vap}) = +40.7 \text{ kJ}$
CH_4	−161 °C	$CH_4(l) = CH_4(g)$	$\Delta H(\text{vap}) = +8.2 \text{ kJ}$

Since we know that the vaporization process for liquid water is endothermic, the thermochemical equation also tells us that a quantity of energy equal to 40.7 kJ *has to be supplied* in order to convert one mole of H_2O molecules in the liquid state, at 100 °C and 1 atmosphere pressure, completely to steam.

▪ How much energy would be required to completely vaporize two moles of H_2O molecules in the liquid state under the same conditions?

■ Common experience tells us that it requires more energy to boil a full kettle than one which is half-full. In these terms, the enthalpy of vaporization must also depend on the quantity of water to be vaporized. Thus, to vaporize two moles of H_2O molecules in the liquid state at 100 °C and 1 atmosphere pressure will require 40.7 kJ × 2 = 81.4 kJ. In terms of a thermochemical equation, we could write

$2H_2O(l) = 2H_2O(g)$ $\Delta H(\text{vap}) = +81.4 \text{ kJ}$

To determine the enthalpy change when a given mass of water is vaporized requires the calculation to be carried out in two steps. First, it is necessary to determine the mass in grams of one mole of H_2O molecules.

▪ How would you calculate this?

■ The relative molecular mass of H_2O is given by:

relative molecular mass (H_2O) =
2 × relative atomic mass (*H*) + relative atomic mass (O).

Given that the relative atomic mass (*H*) = 1.01 and the relative atomic mass (O) = 16.0, and taking into account the 'rules-of-thumb' for calculations described in subsection 4.2.2, then relative molecular mass (H_2O) = 18.0. Thus, 1.00 mol H_2O molecules has a mass of 18.0 g.

▪ Using the result above, see if you can now calculate the enthalpy change when 1.00 g of liquid water is vaporized at 100 °C.

■ If the mass of 1.00 mol of H_2O molecules is 18.0 g, then it follows that

$$1.00 \text{ g } H_2O = \frac{1.00}{18.0} \text{ mol } H_2O \text{ molecules}$$

The energy required to vaporize 1.00 g of liquid water must therefore be

$$\frac{1.00}{18.0} \times 40.7 \text{ kJ} = 2.26 \text{ kJ}$$

Hence, for this particular quantity of water, we can write
$\Delta H(\text{vap}) = +2.26 \text{ kJ}$.

The conversion of a gas to a liquid, which is the reverse of vaporization, is called condensation. If vaporization is an endothermic process, then it follows that condensation must be an exothermic process.

▩   How would you write a thermochemical equation to represent the condensation of one mole of water molecules in the gaseous state at 100 °C, assuming the change occurs at 1 atmosphere pressure?

■   The thermochemical equation will be

$$H_2O(g) = H_2O(l) \quad \Delta H \text{(condensation)} = -40.7 \text{ kJ}$$

Question 15 Given this description of condensation, try to suggest why steam causes such severe scalding when it comes into contact with the skin.

An enthalpy change also accompanies the change of state when a solid melts to give a liquid. Once again, water provides a very good example (Figure 6.2). At normal atmospheric pressure, ice melts at 0 °C to give liquid water at the same temperature.

▩   In brief, and thinking back to Book 1, what would you say were important features of the structure of ice?

■   As described in Book 1, ice has a very ordered structure with each oxygen atom of each water molecule being involved in two hydrogen bonds. This bonding results in a somewhat open structure and also accounts for the fact that ice is less dense than liquid water at 0 °C. (A stereoscopic model for the structure of ice, taken from Book 1, is given in Figure 6.3.)

Figure 6.2
Melting ice.

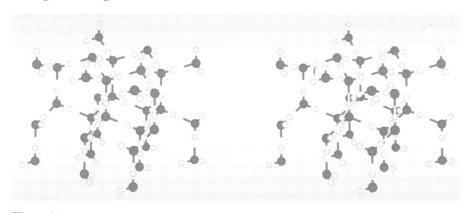

Figure 6.3
A stereoscopic model for the structure of ice, as discussed in Book 1.

Activity 6 Ice and liquid water

This Activity is designed to help you summarize for yourself what you have learnt about what happens when there is a change from one state to another. It uses as an example the change of state when liquid water freezes at 0 °C, and under normal atmospheric pressure, to give ice.

(a) How would you write an equation to describe the change of state when ice melts to form liquid water at 0 °C?

(b) Experimentally, it is found that the energy required to melt one mole of H_2O molecules in the form of ice at 0 °C is 6.01 kJ. How would you write the thermochemical equation for the change of state described in (a)?

(c) How would you write the thermochemical equation to describe the change of state when one mole of H_2O molecules in the form of liquid water freezes to form ice at 0 °C?

(d) What would be the enthalpy change when 1.00 g of water melts at 0 °C?

(e) Briefly, suggest in molecular terms why you think that the enthalpy change of melting is significantly less than the enthalpy change of vaporization for water.

Figure 6.4
Freeze drying is a process that can be applied to a number of products in the food industry; for instance, it is used in making instant coffee. The photograph shows trays of frozen (at −40 °C) coffee extract in a freeze-drying chamber. Under the effect of low pressure and controlled heating, the water in the ice crystals passes directly into the gaseous state without liquid water being formed. Left behind are the coffee granules with which we are so familiar.

On a cold frosty morning, you may have noticed that sometimes the frost seems to vanish and yet no liquid is left behind. You might also be aware that a common procedure used in the food industry relies on a process that has similarities with this natural phenomenon (Figure 6.4). Underlying these observations is the fact that ice can undergo sublimation.

▨ How would you define sublimation?

▪ It is the direct change of state from a solid to a gas without the formation of a liquid.

Sublimation is always an endothermic process and in the case of ice can be represented by the thermochemical equation

$$H_2O(s) = H_2O(g) \qquad \Delta H \text{(sublimation)} = +51.1 \text{ kJ}$$

The relatively large enthalpy of sublimation is a reflection of the fact that water molecules are 'securely locked together' in ice, whereas in the gas phase they can move about relatively freely.

Box 6.1 Evaporation

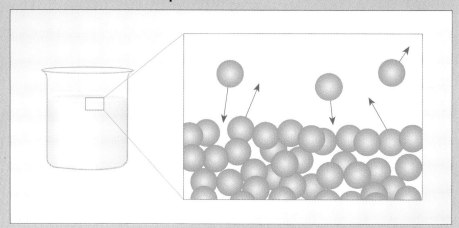

Figure 6.5
A molecular view of the surface of a liquid.

If any liquid that is in contact with its surroundings (like a spill of water on the kitchen floor) is left, then we know that, in time, it will all disappear. The process responsible, as mentioned in Book 1, is evaporation.

The space above any liquid substance always contains vapour consisting of molecules of that substance. For some liquids, the presence of this vapour is very obvious – as you saw, for example, with bromine liquid in Chapter 2. If we could look, at the molecular level, at the surface of a liquid, it would appear something like that shown in Figure 6.5. Some molecules in the liquid will obtain sufficient energy, perhaps by advantageous collisions with other molecules, to be able to escape from the surface of the liquid. By contrast, other molecules which are already in the vapour will come close to the surface and be captured once again. So, both vaporization and condensation are always occurring at the surface of a liquid.

For a liquid that is left 'open to the air', the vapour will be quite free to move away. In molecular terms, the chances of molecules being recaptured at the liquid surface will therefore be diminished and, hence, evaporation takes place.

Forced evaporation, such as happens when a liquid is placed in a draught or on a surface which is warm, tends to have a noticeable cooling effect on the surface. This is because the process of vaporization is taking place and its endothermic character draws in heat from the surroundings including the surface. Indeed, one of the strategies used by our bodies to keep cool is to rely on the evaporation of perspiration.

The evaporation of water has immense global consequences which are linked primarily to its anomalously high enthalpy of vaporization. Just over 20% of the total solar energy intercepted by the Earth goes into driving the vast mechanism of the hydrological cycle, i.e. the evaporation and precipitation (rain and snow) cycle (Figure 6.6). The magnitude of the enthalpy of vaporization is just about right to maintain the generally low humidity of the atmosphere and ensure the stability of the oceans. It is evaporation of ocean water in the tropics that keeps the equatorial regions from being as hot as they otherwise would be. Estimates suggest that the amount of ocean water that evaporates each year at the equator is equivalent to losing a depth of 2.6 metres from the ocean, but it is replenished by precipitation.

Figure 6.6
A typical rainy day. Underlying this familiar picture is the complex working of the hydrological cycle. Water vapour from any source – oceans, seas, lakes and rivers – once formed is transported by winds until it finally condenses and returns to Earth as rain or snow. In some areas of the world, very regular seasonal patterns may be established as, for example, with monsoon rain; in others, there may be no pattern at all – as we know so well.

Question 16 You have, so far, met a number of terms that describe changes from one state to another. As revision, see if you can complete the middle column of Table 6.2 with the term used to describe each change.

Table 6.2 Changes from one state to another.

Initial state	Change	Final state
solid	*melting*	liquid
solid	*sublimation*	gas
liquid	*freezing*	solid
liquid	*evaporation* *vaporization*	gas
gas	*condensation*	liquid
gas	*deposition*	solid

6.2 A good fuel guide

Figure 6.7 shows the use of primary energy sources which were commercially traded in the UK in 1993 and confirms that most of our daily energy needs are provided by coal, oil and natural gas.

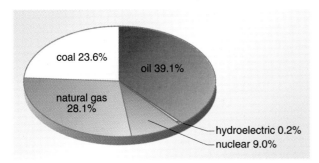

Figure 6.7
Primary energy sources which were commercially traded in the UK during 1993.

A key characteristic of these fuels is that their combustion produces heat which, for most conventional applications, can then be used either directly or be transformed into another major source of energy: electricity. The concept of fuels and the energy they supply has been referred to in descriptive terms as a 'fuel package'; both the fuel, and the oxygen in which it burns, are equally important in this package. Also, it is usually the case that some form of ignition process is also required. Overall in combustion a 'reshuffling' of matter takes place at the molecular level and the internal energy changes involved are such that energy is released. But having said this, a natural question arises as to which fuel is best selected for a particular application. What criteria would you suggest are important? Try to make a quick list before reading on.

Various criteria can be gathered under different headings as below (although no particular order of importance is implied).

Economic and political circumstances of the day. The costs associated with exploration and extraction cannot be ignored; indeed, you saw in Chapter 4 that these factors contribute substantially to the uncertainties inherent in the term 'proved reserves'. A glance back to Figure 4.8 should convince you of the importance of political considerations.

Environmental issues. The effects of pollution have to be considered. This means that the gases produced in the combustion process must not be toxic or damaging to the environment. So, potential fuels must contain no, (or limited amounts of) elements like sulfur or nitrogen which give harmful gaseous oxides on combustion. The production of 'greenhouse gases', in particular carbon dioxide, is an area of considerable concern and complexity.

Technicalities. These relate, for example, to how easily a fuel can be burned or how smoothly it detonates. The latter is very important for car engines.

Storage and transportation. The ease and expense of both of these has to be taken into account. Solid fuels, like coal, are more easily stored than gases or liquid fuels. But the latter are easily transported from place to place using pipelines. Liquid fuels offer the easiest transfer and handling characteristics for mobile energy sources. Any fuel requiring refrigeration will have very restricted use.

Missing from the list is one criterion you might have thought of immediately: the energy content of the fuel. Or, in other words, the amount of energy that can be gained from a given quantity – mass or volume – of the fuel. It is this criterion that will be the focus of our attention in the next few pages.

To illustrate how we can compare different fuels on an 'energy basis', it is useful to consider first a specific example; for convenience, we shall take methane. This gas will burn *completely* in a plentiful supply of oxygen to give carbon dioxide and water vapour. The balanced chemical equation for the combustion reaction is

$$CH_4(g) + 2O_2(g) = CO_2(g) + 2H_2O(g)$$

It's worthwhile re-emphasizing what we understand by this equation. In *molar terms* it tells us that for every one mole of CH_4 molecules (i.e. 6.02×10^{23} molecules) that react, two moles of O_2 molecules also react. Since all of the methane is used, the product mixture consists of one mole of CO_2 molecules and two moles of H_2O molecules; in summary:

1 mol CH_4 molecules + 2 mol O_2 molecules =
                    1 mol CO_2 molecules + 2 mol H_2O molecules

As indicated by the physical state label 'g', all of the substances involved in the reaction are in the gas phase, including the water that is formed. In practice, and depending on circumstances, water could be formed as either a gas or a liquid; for convenience, we shall always assume that it is in the former state.

In a comparable manner to that in which we wrote thermochemical equations to describe changes of state, we can also adopt a similar approach for chemical transformations. The combustion of methane gas is an exothermic process and experimentally it is found that the enthalpy change of combustion when a quantity of methane equivalent to one mole of CH_4 molecules is burned fully in oxygen is given by $\Delta H = -804$ kJ.

■   How would you write a thermochemical equation for the complete combustion of methane gas?

■   The thermochemical equation will be

$$CH_4(g) + 2O_2(g) = CO_2(g) + 2H_2O(g) \qquad \Delta H = -804 \text{ kJ}$$

The thermochemical equation tells us that the heat released by the reaction when it occurs at constant pressure, say in the open atmosphere, corresponds to an energy transfer of 804 kJ. If we viewed the reaction as also occurring at constant temperature, then this is the quantity of energy that would be transferred directly to the surroundings. Of course, in practice, the situation is very different and initially a good deal of the energy released goes into raising the temperature of the reaction mixture.

A good way to compare different fuels is to look at their thermochemical equations – in effect, the enthalpy change provides information about the 'heat available'. To facilitate the comparison, it is easiest to write the balanced chemical equations *so that only one mole of fuel molecules is involved*. This approach also helps in calculating the amount of heat that will be released by a given mass of a particular fuel. It is sensible to standardize on a mass that gives 'reasonably sized' numbers for all fuels: it turns out that one gram is fine. A 'recipe' can be developed for calculation purposes. This is described in Box 6.2.

Box 6.2 A method for calculating the heat released by one gram of a given fuel

There are five steps in the method (opposite).

Note that the first step in this recipe may mean that the 'number' which appears in front of $O_2(g)$ in the balanced chemical equation for the combustion involves 'a half'. If you look forward to Table 6.3, you will see that this is the case for the combustion of octane, C_8H_{18}. In a molar sense, we simply interpret '$12\frac{1}{2}O_2(g)$' as 'twelve-and-a-half moles of oxygen molecules'.

1 We shall assume that the balanced chemical equation is always written – as already described – so that one mole of fuel molecules is involved; in words, one mole of fuel molecules reacts *fully* with oxygen to give products.

2 One mole of fuel molecules can be converted to a mass of fuel by determining the relative molecular mass of the fuel and then expressing this quantity in grams.

3 The enthalpy change given in the thermochemical equation will correspond to the change when this mass of fuel is fully burned.

4 The enthalpy change when one gram of fuel is fully burned will be smaller and can be calculated by taking 'the enthalpy change given in the thermochemical equation and dividing by the relative molecular mass of the fuel expressed in grams'.

5 The heat released when one gram of fuel is burned can then be stated.

Activity 7 Carrying out the recipe:
Calculating the heat released when one gram of methane is fully burned in air

Reading a recipe is one thing, carrying it out successfully is another. For practice, see if you can use the method described in Box 6.2 to determine the enthalpy change when 1.00 g of methane is fully burned in oxygen. If you find difficulty, then take the time to work carefully through the comments for this Activity.

We are now in a position to produce our 'good fuel guide', although we shall not ask you to carry out any further calculations. The version we shall concentrate on is given in Table 6.3. As you can see, this Table compares a selection of fuels (or 'simple' compounds which are related to them), which are used in our homes, motor cars, industry, our own bodies and even in rockets. (You are not expected to remember all of the detail in this Table.)

Table 6.3 Combustion of typical fuels.

Fuel	Thermochemical equation	Heat released per gram of fuel[a]
carbon, C	$C(s) + O_2(g) = CO_2(g)$ $\Delta H = -394\,kJ$	33 kJ
methane, CH_4 (as in natural gas)	$CH_4(g) + 2O_2(g) = CO_2(g) + 2H_2O(g)$ $\Delta H = -804\,kJ$	50 kJ
octane, C_8H_{18} (as in petrol)	$C_8H_{18}(l) + 12\tfrac{1}{2}O_2(g) = 8CO_2(g) + 9H_2O(g)$ $\Delta H = -5\,120\,kJ$	45 kJ
methanol, CH_3OH	$CH_3OH(l) + \tfrac{3}{2}O_2(g) = CO_2(g) + 2H_2O(g)$ $\Delta H = -640\,kJ$	20 kJ
ethanol, C_2H_5OH	$C_2H_5OH(l) + 3O_2(g) = 2CO_2(g) + 3H_2O(g)$ $\Delta H = -1\,238\,kJ$	27 kJ
carbohydrates (sugars and starches)	products are $CO_2(g)$ and $H_2O(l)$	an average value of 17 kJ
animal fats	products are $CO_2(g)$ and $H_2O(l)$	variable, 40 kJ a typical value
hydrogen, H_2	$H_2(g) + \tfrac{1}{2}O_2(g) = H_2O(g)$ $\Delta H = -243\,kJ$	120 kJ

[a] These values are to be used for comparative purposes and so are simply rounded to the nearest whole number.

The thermochemical equations given in the table are of basic interest but it is more than likely that your attention focused on the column giving figures for the 'heat released per gram of fuel'. It is these figures that provide practical information and act as starting points for wider-ranging discussions, as you will see shortly.

> **Question 17** A South Wales anthracite was quoted as having a 'heat value' of 9.88 kWh per kilogram. In Table 6.3, the heat released per gram of solid carbon is given as 33 kJ. How do these two figures compare?

It is important to notice that all of the fuels in Table 6.3, with the exception of hydrogen, produce carbon dioxide as one of the products. These anthropogenic ('created by humans') sources of carbon dioxide are known to be affecting the natural concentration of this gas in the atmosphere (Figure 6.8).

The best estimates for total worldwide emissions of carbon dioxide in 1990 were in the region of 30 billion tonnes. Of this figure, a significant proportion (70%) was attributed to the burning of fossil fuels with the remainder being accounted for by changes in land usage. The latter, which includes the effects of deforestation and the spread of farming, is very difficult to quantify but there is a consensus that, overall, the changes

result in a *net* emission of carbon dioxide. Cement production also makes a contribution: it is generally estimated that the heating of limestone in the manufacturing process drives off some 0.5 billion tonnes of carbon dioxide annually.

The total atmospheric concentration of carbon dioxide currently amounts to about 2 700 billion tonnes. This 'store of carbon', however, is by no means static because it is intimately involved in the vast and complex cycle known as the carbon cycle. This cycle involves natural exchanges of 'carbon' between the atmosphere, the oceans and living materials. On an annual basis, anthropogenic emissions of carbon dioxide act as a relatively small perturbation within this vast cycle, but in the longer term this may not be the case. In particular, the concern is whether the *natural* greenhouse effect, which is so vital to life on Earth, will be enhanced and result in 'global warming'; this is discussed in more detail in Chapter 9. For now, it is important to recognize that carbon dioxide is a gas that is very much involved in environmental issues.

Figure 6.8
The oil-well fires in Kuwait during 1991 released an amount of CO_2 equivalent to about 1% of that expected to be released by the burning of fossil fuels on an annual, and worldwide, basis. The Al Burgan field is seen here in which over 360 oil wells were ablaze at one time.

6.2.1 Hydrogen as a fuel?

The last entry in Table 6.3 – that for hydrogen – demonstrates clearly that on a 'mass-for-mass basis', hydrogen is a very good fuel. In principle, it could be transported efficiently through pipelines to where energy was needed; and, for example, in technical terms it would not be difficult to burn in cookers, industrial furnaces, motor cars or aeroplanes. Such a scenario has led to speculation about a possible future 'hydrogen-based economy' (Figure 6.9). This would release us, in principle, from our dependence on, and the environmental problems associated with, fossil fuels. Furthermore, the combustion of hydrogen in pure oxygen is 'clean' in that the principal product is just water. The combustion in air is not as clean because at the temperature involved nitrogen in air will also react with oxygen to give harmful nitrogen oxide gases: these are popularly known as NO_x (spoken as 'nox'). Whether a hydrogen economy will ever be established remains to be seen, but it seems unlikely in the foreseeable future.

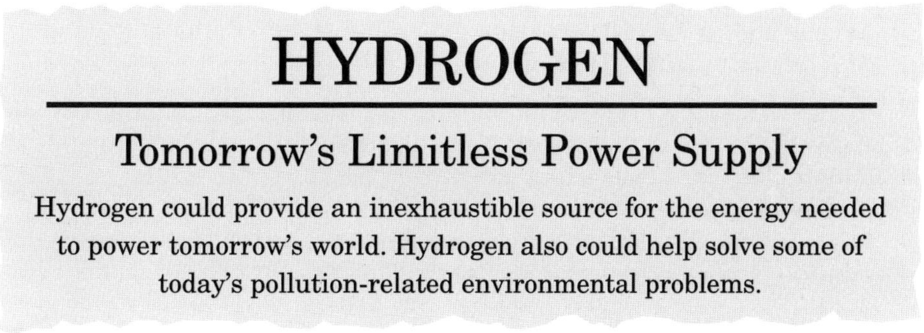

Figure 6.9
The headline from a typical article.

HYGROGEN

Tomorrow's Limitless Power Supply

Hydrogen could provide an inexhaustible source for the energy needed to power tomorrow's world. Hydrogen also could help solve some of today's pollution-related environmental problems.

One major problem is that hydrogen is not a natural energy source. It is the most abundant element in the Universe, but because it escapes from our atmosphere with relative ease there is little of it on Earth. We have to treat it as a synthetic fuel and so it has to be manufactured – and this requires energy. One possible way of achieving the production of hydrogen is by electrolysis. As you may recall from Book 1, this is a process which involves passing an electric current through a liquid. The result for water is that hydrogen as well as oxygen is produced, as illustrated in Figure 6.10.

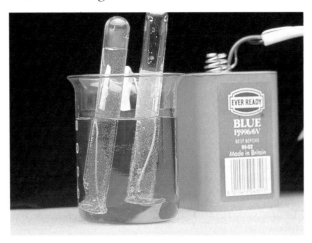

Figure 6.10
This simple experimental arrangement shows how water may be electrolysed. The two wires from the battery that enter into the water are stripped at their ends and act as electrodes. The electric current passing through water decomposes it into hydrogen and oxygen gas according to the chemical equation:
$2H_2O(l) = 2H_2(g) + O_2(g)$.
The bubbles due to these gases are clearly seen coming from the electrodes.

Estimates suggest that about half of the energy used in electrolysis can be recovered from the hydrogen gas liberated. Realistically, for a society to adopt a hydrogen-based economy centred on electrolysis, it must have large surplus amounts of electricity. Furthermore, this electricity should be generated by methods that do not harm the environment. Such a society would then be able to store its excess electrical energy – as chemical energy in hydrogen gas – and so replace present natural gas and other hydrocarbon fuels as stores of energy.

On a smaller scale, the idea of 'storing' energy in hydrogen gas may be more feasible. As one author argues, 'if a cheap and simple electrolysis "kit" could be produced, the intermittent and unpredictable output of a wind-powered generator could be stored and the surplus power of a solar-powered generator could be slowly accumulated during the summer and used in winter'. Electrolysis may turn out to be not the best method of producing hydrogen gas from water. Laboratory-based experiments have shown, for example, that water can be decomposed in the presence of sunlight as long as special (but often expensive) materials that make the process more efficient are present.

Hydrogen as a fuel is not without drawbacks. The airships, or dirigibles, of the early 1900s, e.g. the rigid type constructed by the Zeppelin company, used hydrogen gas, not as a fuel directly but because of its 'lifting properties'. Between 1910 and 1914, Zeppelins were in extensive use and proved a remarkably safe passenger service. But tragically, following the resumption of passenger services after the First World War, disaster

Figure 6.11
The hydrogen-filled airship *Hindenburg* was destroyed at its moorings at Lakehurst, New Jersey, in 1937.

followed disaster. The public's realization that a spark or flame could lead to a sudden conflagration of an oxygen and hydrogen mixture gave hydrogen-filled airships a bad reputation from which they never recovered (Figure 6.11).

The first British rigid-type airship to fly was *R9* in November 1916. It contained just over 25 000 m³ of hydrogen gas with a mass of less than 2.5 tonnes. (The same volume of air would have a mass of roughly 36 tonnes.) Hydrogen is a gas that is certainly 'lighter than air' and, indeed, it has the lowest density of any known gas.

■ What effect will the density of hydrogen gas have on its suitability as a fuel?

■ A given *volume* of the gas will have only a relatively small mass and so the enthalpy change for combustion for this volume will be small.

On a 'volume-for-volume' basis, it turns out that the enthalpy change for hydrogen gas on combustion is only one-third of that for natural gas. In these terms, it is not such a good fuel, but it is the environmental credentials of hydrogen that give it a potential advantage.

Hydrogen is a liquid between a narrow range of very low temperatures (−253°C to −259°C), and so the costs of refrigeration and storage of hydrogen as a liquid effectively prohibit it from becoming a more widely available fuel. Furthermore, the density of the liquid is still relatively small – less than one-tenth that of water – and so any storage tanks would have to be large. There is, however, one very specialized use of liquid hydrogen as a fuel which you may be aware of already. It is used in combination with liquid oxygen in the main engines of large rockets, including that of the US Space Shuttle (Figure 6.12). (The latter also needs booster rockets which are based upon a solid fuel.) The energy packed per unit mass in liquid hydrogen is important but, in addition, the explosive nature of the reaction provides enough thrust to take a rocket to orbital velocity (about 7 600 metres per second, which is much faster than the average speed of an oxygen molecule at room temperature.)

Figure 6.12
The US Space Shuttle *Discovery* at lift-off. The large fuel tank contains liquid oxygen and liquid hydrogen. For a typical mission, about 1.4 × 10⁶ litres of liquid hydrogen are needed. The solid fuel booster rockets are seen either side of the main fuel tank.

A possible future means of storing hydrogen gas is within a metal or metal alloy. This illustrated in Figure 6.13.

> **Question 18** Suppose you were asked to summarize the prospects for hydrogen as a fuel in a simple table labelled 'advantages' and 'disadvantages'. Furthermore, each entry in the table should be made using only a minimum of words. What would you put?

6.2.2 Petrol: additives and alternatives

A fuel that many people – some would argue too many – use every day is petrol, or as it is called in other countries 'gas' or 'gasoline'. Technically, it is a refined petroleum product that provides a suitable fuel for the internal combustion engines of many modern motor vehicles. The technical processes by which petrol is obtained, or refined, from crude oil are complex. Refiners also have their own special methods, plus a little 'witchcraft', to offer savings in production costs, and to cope with the variability in the crude oil that is handled. Only the briefest of sketches of the way in which petrol is produced will be given here.

Figure 6.13
It is of current interest that some metals and alloys will absorb hydrogen gas in such a way that in a given volume of metal there are effectively more hydrogen molecules than in a similar volume of its own liquid. This offers the prospect of safe storage for hydrogen which can then be released, when needed, by controlled heating. This technology is used in the prototype car above, which has a range of nearly 200 miles. The hydrogen is not burned but is used in a device called a fuel cell to provide electrical power.

The first important step in the refinement process is called **fractional distillation**. Underlying this process is the general observation that the normal boiling temperatures of simple straight-chain alkanes tend to increase as the number of carbon atoms in the structure increases. For example, at ordinary pressures, butane, C_4H_{10}, and heptane, C_7H_{16}, boil at just below the freezing and boiling points of water, respectively. Rather than use the names of the alkanes, it is common practice when discussing refining to refer to them by the number of carbon atoms they contain; thus butane is C_4 and heptane is C_7 (Figure 6.14). A schematic drawing of a fractional distillation tower is shown in Figure 6.15.

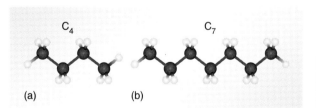

(a)   (b)

Figure 6.14
(a) Butane can be labelled C_4 and (b) heptane can be labelled C_7.

From crude oil, the fractional distillation process produces fractions which have distinct boiling temperature ranges. Each fraction contains a variety of hydrocarbons each of which will contain a distinct number of carbon atoms. The range of carbon atom numbers in a fraction, *irrespective of the chemical structures of the hydrocarbons present*, provides a simple means of classification. Thus, for example, kerosene is a fraction with a boiling temperature range of 175–320 °C, or a carbon atom number range of

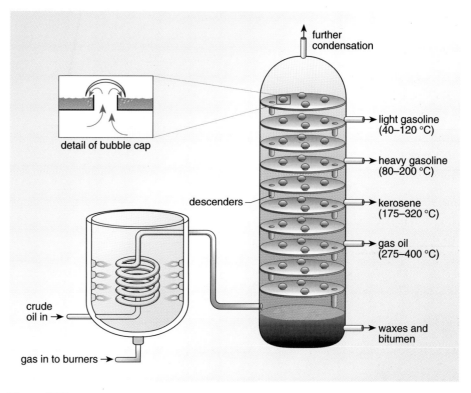

Figure 6.15
In the fractional distillation process, crude oil is heated and fed into the fractionating tower. Typically, this can be as high as 50 m and the temperature within it decreases with increasing height. The hot liquid from the crude oil is drawn off as 'bottoms' and is used to provide waxes and bitumens. The hot vapour rises within the tower and condenses at various points. This process is assisted by the tower being divided into sections by perforated trays with each perforation having a 'bubble-cap'. It is in these bubble-caps that various components of the rising vapour condense to form a liquid which fills the trays. Some of this liquid is fed back by descenders to be re-vaporized at lower points in the tower: the remainder is drawn off as fractions. In the hot, lower parts of the tower, only the hydrocarbons with the highest boiling temperatures will condense; the components with the lower boiling temperature will remain in the vapour phase longer and only condense in the cooler, higher parts of the tower. The Figure shows some of the important fractions that are drawn off, classified according to their boiling temperature ranges.

typically C_{12} to C_{16}. The light and heavy gasoline fractions, which are often remixed, are generally referred to as '*straight-run gasoline*' in the USA, or '*straight-run petrol*' in the UK. They contain hydrocarbons with 5–10 carbon atoms, i.e. C_5–C_{10}, and these hydrocarbons are chiefly of the straight-chain variety although for each carbon number there will be a whole variety of isomers present.

Before the motor car became popular in the early 1900s (Henry Ford built his first car in 1896 and the first Rolls-Royce Silver Ghost was sold in 1907), most crude oil was refined to give kerosene and this liquid was used mainly for heating and lighting. The early motor car engines were designed to burn a form of the straight-run gasoline fraction. This mixture evaporates quite readily and it is perhaps for this reason that it became known as *gas*oline. Such a fuel would not be suitable for modern motor vehicles, for reasons which will become apparent.

Figure 6.16
A modern oil refinery.

Modern petroleum refineries use additional processes (more towers and a lot of pipework as shown in Figure 6.16) to increase both the *yield* and the *quality* of petrol from a barrel of oil; since the 1960s, over 40% of a barrel can be converted to 'modern petrol'. Two key chemical processes are involved. One of these is **cracking** in which zeolites play a crucial role. Basically, less valuable, higher boiling temperature fractions are processed in such a way that the long-chain hydrocarbon molecules they contain are split to give shorter-chain hydrocarbons. Some of these will have carbon atom numbers in the straight-run range. For example, a C_{16} hydrocarbon might be converted to C_{10} and C_6 hydrocarbons. In this way, more straight-run petrol is obtained from a barrel of oil: cracking particularly addresses quantity of final product. Quality, which is related to 'suitability' for modern motor vehicles, is enhanced by the process of **reforming**. This process changes the 'chemical mix' of straight-run petrol so that it is no longer dominated by straight-chain hydrocarbons. A lot more branched-chain hydrocarbons are introduced as well as hydrocarbons containing the benzene ring. A typical reforming reaction is shown in Figure 6.17 in which straight-chain heptane is converted to a compound called toluene, which is based on the benzene structure. As for most reforming reactions, the number of carbon atoms in the starting and final products remains the same.

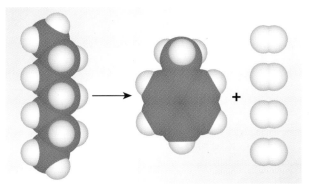

Figure 6.17
A typical reforming reaction. Heptane and toluene are both C_7 molecules. The other product in the reaction is hydrogen gas.

Activity 8
An exercise with unleaded petrol

Figure 6.18
A measuring cylinder containing ordinary unleaded petrol.

The measuring cylinder contains 100 g of ordinary unleaded petrol of volume 133 cm³. Assuming that the information given for octane, C_8H_{18}, in Table 6.3 is typical, make an estimate of the 'energy content' of each litre of petrol that is bought at petrol stations. Do you think that all of this energy is used in driving the wheels of a motor car (that is converted to kinetic energy)?

The final question asked in Activity 8 is an important one. A general 'rule-of-thumb' for the modern motor car is that, of the energy developed during combustion, one-third is lost as heat to the cooling system and one-third is lost as friction, leaving the remaining one-third to propel the vehicle forwards. Matters are worse in the 'stop-and-go' of urban driving where the efficiency of the engine averages only about 15%. There are very fundamental reasons why the maximum efficiency of the internal combustion engine cannot be more than about 55%, but this still leaves room for improvements. Better car design in terms of aerodynamics and mass, computer control of engine functions, and perhaps replacement of steel by ceramics in engines (the latter have better thermal properties), will all give rise to improved efficiency in the future. It is also important to improve the technology to reduce emissions from cars. These include unburned fuel and carbon monoxide, nitrogen oxides (NO_x) and, until recently, lead compounds – all of which are harmful to the environment.

To understand more about the 'quality' of petrol, it is necessary to have some idea of what happens in the cylinder of a modern car engine. A brief outline is given in Box 6.3 overleaf.

If the straight-run gasoline fraction was used in modern car engines, it would give rise to '**knocking**'. This particular foible of the internal combustion engine was being researched as early as 1916 by an engineer called Thomas Midgley (1889–1944) in the United States. It was established that the sound (sometimes called 'pinking' or 'pre-ignition') associated with a knocking engine was metallic rattling of engine components, particularly when the engine was under load. (Modern cars still pink if they are badly tuned or if too high a gear is used on a steep hill). Overall, the effect reduces engine power and ultimately causes damage. Midgley diagnosed the root cause of the problem as being due to uneven combustion of the fuel/air mixture during the ignition stroke. Ideally, the flame front (Figure 6.20), starting from the sparking plug, should accelerate smoothly through the compressed fuel/air mixture. However, the mixture in those parts of the cylinder remote from the sparking plug – known as end-gas – undergoes relatively prolonged heating and compression as the flame front approaches. Knocking arises if this end-gas undergoes premature detonation, so that there are disorderly explosions in the cylinder. Midgley and his team, in an extended 'try it and see' approach involving many compounds, finally found an antidote in the compound lead tetraethyl (Figure 6.21). A small amount of this compound in the fuel soothed the problem; it allowed the fuel/air mixture to burn smoothly through the power stroke. Lead tetraethyl is said to improve the **octane rating** of the fuel.

Box 6.3 Within the cylinder of a modern car engine

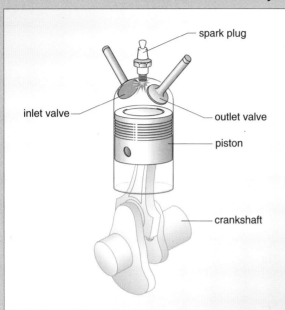

Figure 6.19
A schematic view of one cylinder of an internal combustion engine at the point of detonation of the fuel/air mixture.

The energy to move a car is derived from the combustion of a mixture of petrol and air in the internal combustion engine. The combustion takes place in specially designed cylinders. The operation of the cycle of combustion can be described as follows.

At the top of the cylinder, the inlet valve opens and, at the same time, a mobile piston moves downwards, sucking in a mixture of fuel and air. This is *induction*. When the piston reaches the bottom of the cylinder, the inlet valve closes and the piston begins to move upwards. As it does so, it rapidly compresses the mixture of fuel and air. Hence, this is called *compression*. Just before the moment of maximum compression, a pulse of electricity passes across a gap in the spark plug and this detonates the fuel/air mixture. This leads to *ignition* which involves the downward stroke of the piston propelled by the explosive force of the chemical reaction between the fuel and air. Finally, there is the return upwards of the piston in an *exhaust stroke*, during which the exhaust valve is open and exhaust gases are pushed out. When the piston reaches the top of its stroke for a second time, the exhaust valve closes, the inlet valve opens, and the cycle begins again; i.e. *induction → compression → ignition → exhaust stroke*. The movement of the piston is connected to a crank, and this gives the car its propulsive force through a system of gears and wheels.

For each cycle, the piston makes two strokes up, and two strokes down, and so this kind of engine is referred to as a 'four-stroke' engine.

An important characteristic of the petrol engine is the compression ratio. In technical terms, this is the ratio of the volume in the cylinder at the end of the induction stroke, to the volume in the cylinder at the end of the compression stroke. In practical terms, engines with high compression ratios develop more power and consume less fuel; modern car engines have ratios of, typically, about 9 : 1 or 10 : 1.

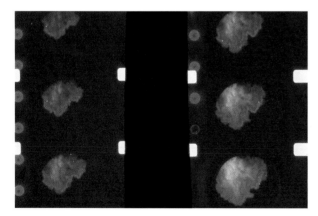

Figure 6.20
The combustion of the fuel/air mixture during the ignition stroke produces a large local temperature rise and, consequently, a flame. The photograph is of six successive frames from a high-speed film.

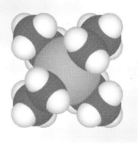

Figure 6.21
Apart from its huge success as an additive, it was later discovered that lead tetraethyl also helped to reduce 'wear and tear' on the valve systems themselves by providing a lubricant of lead oxide.

Box 6.4 Octane rating

An 'octane rating' is an arbitrary scale which reflects the relative knocking properties of a given petrol based on the operation of a standard engine running under specified conditions. These conditions can be standardized to reflect different driving circumstances (e.g. mild or severe), and so give rise to different octane rating scales. Here we shall simply use the term 'octane rating' and take it to refer to mild driving conditions.

Heptane as a fuel causes considerable engine knock in the standard engine and is assigned an octane number of zero.

In contrast, iso-octane (full chemical name 2,2,4-trimethylpentane), which is a branched-chain hydrocarbon, has far smoother combustion qualities in the standard engine; it is therefore assigned an octane number of 100 (Figure 6.22).

To determine the octane rating of a given motor vehicle fuel, it is used in a standard engine and its knocking properties are recorded. The percentage (by volume) of iso-octane that must be blended with heptane to give identical knocking properties is called the 'octane rating'. Thus, if a given petrol has the same knocking characteristics as a mixture of 7% heptane and 93% iso-octane, then it has an octane rating of 93. If a fuel performs better than neat iso-octane then it is given an octane number greater than 100. If it performs worse than neat heptane, then it is given a negative octane number which is the case for octane itself.

The greater the octane rating of a fuel then the greater is its ability to withstand high pressures and temperatures and yet still facilitate smooth combustion. This is important for modern car engines which have high compression ratios.

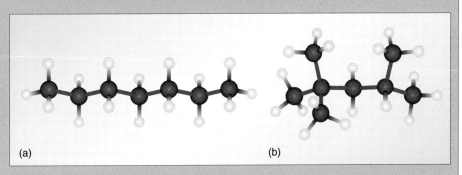

Figure 6.22
(a) Heptane and (b) iso-octane.

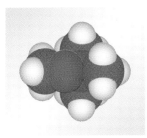

Figure 6.23
An important octane enhancer is MTBE which is an abbreviation of the common name 'methyl tertiary butyl ether'. It is especially effective as it has an octane number itself of greater than 100. It is one of a class of enhancers known as oxygenates because oxygen is contained in the molecular structure.

The octane rating of straight-run petrol usually lies in the range 50–55. While adequate for the relatively low-compression ratio (typically 4 : 1) cars of the 1920s, this fuel would cause severe knocking in modern car engines. The latter need a fuel with an octane rating of greater than 90, with 4-star petrol being characterized by a rating of at least 97. During the 1970s, petroleum refineries increased the rating of the straight-run fraction to just below 90 by a combination of cracking and reforming. Final upgrading was achieved by adding lead tetraethyl; as little as three grams per gallon improves the rating by 10 to 15. But this was prior to the acceptance of the medical dangers of car emissions as sources of airborne lead compounds. Environmental pressures subsequently resulted in policies to introduce 'green' or 'lead-free' petrol. Oil companies now sell lead-free petrol with an octane rating of greater than 95. This has been achieved in various ways. For example, in the UK this has been partly by further refining, and partly by blending in more environmentally acceptable 'octane enhancers' (Figure 6.23).

Relatively minor modifications to internal combustion engines will allow them to burn neat ethanol (octane number 108): indeed, this fuel was used to supplement petrol supplies in the First and Second World Wars.

Currently, Brazil has the world's largest programme for using ethanol as a fuel for cars. It was launched in 1975 and by the late 1980s about 4.5 million alcohol-powered cars were on the road. The ethanol was produced from sugar cane, a fast-growing tropical plant (Figure 6.24), by the traditional methods of fermentation followed by distillation. In these terms, ethanol is a liquid biomass fuel where the term 'biomass' is a collective term for naturally occurring materials such as plants and trees. A problem with the programme is cost because the alcohol is not competitive with conventional petrol at the oil prices pertaining since the mid-1980s. The programme still continues but is not without its critics.

Methanol is also a contender as a motor vehicle fuel. For example, a 1989 publication presented 'The case for methanol' with this abstract:
'*The authors maintain that a move to pure methanol fuel would reduce vehicular emissions of hydrocarbons and greenhouse gases and could lessen US dependence on foreign energy sources*'. The idea of methanol-powered vehicles is not new; the technology for burning it in car engines, particularly racing cars (Figure 6.25), has existed for a number of years.

Figure 6.24
A growing fuel supply?

Figure 6.25
IndyCars use methanol as a fuel.

Question 19 Can you suggest what types of criticism might be levelled at the proponents of methanol as a vehicle fuel?

Question 20 Without looking back at the discussion, see if you can remember some of the strategies that can be used to produce lead-free, but knock-resistant, petrol.

6.2.3 Plants, animals and energy storage

Under the action of sunlight, green vegetation absorbs carbon dioxide and releases oxygen. This natural process was first identified as early as 1771 by Joseph Priestley, an English Presbyterian minister and pioneer in the chemistry of gases. We now call the process 'photosynthesis', literally meaning 'putting together by light', and it is certainly the most widespread

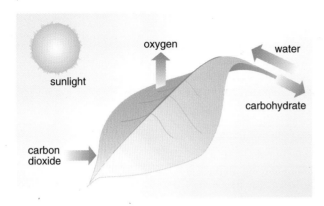

Figure 6.26
A schematic view of photosynthesis.

of all chemical reactions that occur in the Earth's biosphere. Each year, colossal amounts of carbon dioxide are captured from the atmosphere by plants and combined with water, drawn from the soil, to produce carbohydrates; oxygen gas is also released in the process. Photosynthesis is driven by energy provided by sunlight which is absorbed by pigments, chlorophyll being the most dominant, in plant leaves. Chlorophyll is also the substance that takes part in the conversion process and so there are mechanisms by which it can receive the energy collected by other pigments. It is interesting to reflect that the fossil fuels we now mine are essentially no more than stores of the Sun's energy of long ago; what's more, the global photosynthetic process replenishes our atmosphere with oxygen. A schematic view of photosynthesis is shown in Figure 6.26.

In plants, photosynthesis produces glucose which then, in the 'elegant but economical way of nature', is further converted into two different types of material.

▨ From Part 1 of this Book, can you recall the structure of cellulose?

▪ It is a natural polymer formed by linking many hundreds of glucose molecules together. Figure 6.27 shows a simplified representation of a cellulose chain.

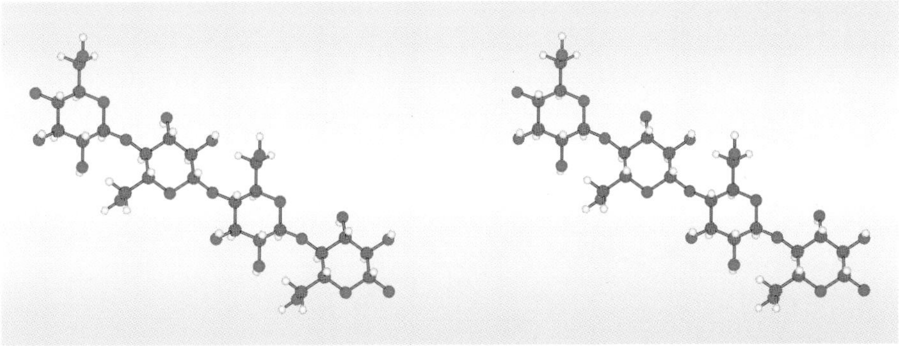

Figure 6.27
A simplified stereoscopic representation of a cellulose chain.

In cellulose, the glucose molecules link together in such a way that the chains are flat and ribbon-like. Hydrogen bonding between the chains results in a fibrous material; for example, cotton fibres consist almost entirely of cellulose. Cellulose functions as a structural component in plants. It cannot be digested by either humans or carnivorous animals.

The other material formed in photosynthesis is starch, which is made up of two natural polymers: amylose (soluble in hot water) and amylopectin (insoluble). The structures of both of these components are reasonably complex, but are essentially different from that of cellulose in that the polymer molecules have a more branched structure. Starch acts an 'energy-storage system' for the plant.

▧   Looking back to Table 6.3, what comments would you make about this
    energy-storage system?

■   The energy released on the oxidation of one gram of starch (no flames
    are involved in a plant!) is in magnitude the smallest entry in the table.
    So on a mass basis, starch would not seem to be the best material to use
    for storing energy.

A chemical reason for the relatively small energy release is that starch
already contains a reasonably high proportion of oxygen and so, in a
sense, is already partially oxidized. Thus, in comparative terms, a smaller
energy release might be expected when a given amount of starch is
completely oxidized to give carbon dioxide and water. But this still begs
the question, 'Why do plants use starch to store energy?' One factor that
probably plays a role is that the chemistry involved in assembling starch
from glucose, and carrying out the reverse, is not too complex – at least
not for the natural processes that occur in the plants. This means that it is
not too difficult to get energy in, or energy out. It might therefore be that
there is no advantage to a *stationary* plant in having a lightweight, energy-
rich, fuel that in chemical terms would be difficult to handle.

Animals, including humans, have two different types of energy store. For
long-term storage, fats are used. A typical fat molecule is shown in Figure
6.28. The details of the structure need not concern us too much except to
notice that a large proportion of the molecule is constructed from
$-CH_2-CH_2-CH_2-$ chains. In this sense, there are similarities to some of
the types of hydrocarbon molecule found in petrol.

▧   What does Table 6.3 tell you about fats as energy-storage systems?

■   On a mass-for-mass basis, fats are more 'energy-rich' than starch and,
    in fact, are not too dissimilar to natural gas and petrol in this particular
    respect.

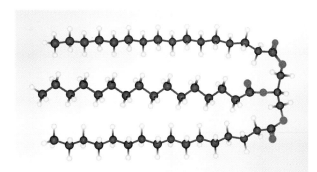

Figure 6.28
A typical fat molecule.

It is tempting, without delving too deeply into evolutionary details, to
simply suggest that nature uses a lightweight fuel for long-term energy
storage for animals which must move about and carry their fuel supplies
with them (Figure 6.29). It is also interesting to note that animals also cater
for their 'rapid-access energy' needs. For this purpose, they have their own
'animal starch' – glycogen – again a polymer of glucose, which is carried,
ready for action, in muscles and the liver.

Figure 6.29
It is generally thought that animals, when unhindered, travel at their most comfortable speed. For kangaroos, this appears to be 20–25 kilometres per hour. The hopping motion of kangaroos is particularly interesting because the energetic costs do not vary over quite a wide range of hopping speeds.

Summary of Chapter 6

This Chapter has focused on two important examples of 'enthalpy changes in practice'; these are:

● enthalpy changes associated with a change of state, and

● enthalpy changes associated with the use of fuels.

In the former, water was taken as the main example and in the latter attention was focused on developing and discussing a 'good fuel guide'.

Changes of state can be represented by equations and can be exothermic or endothermic. For example, the vaporization of liquid water to steam at 100 °C is endothermic while the freezing of liquid water to ice at 0 °C is exothermic. A molecular picture is useful for rationalizing these observations. That hydrogen-bonding is so important in determining the properties of water is emphasized by the fact that liquid water has the largest enthalpy of vaporization, measured per gram, of any known liquid.

A thermochemical equation can be used to describe the enthalpy change when one mole of molecules of a substance undergoes a change of state at a given temperature and constant pressure. Activity 6 was designed to help you summarize for yourself what happens when a substance undergoes a change from one state to another. It used the change of state from liquid water to ice at 0 °C and 1 atmosphere pressure, and *vice versa*, as the example. Sublimation is always an endothermic process.

Any liquid that is left in contact with its surroundings will evaporate. Forced evaporation occurs when a liquid is warmed or placed in a draught. The evaporation of water has immense global consequences.

Criteria for a good fuel can be considered under the headings: economic and political circumstances of the day, environmental issues, technicalities, storage and transportation, and 'energy content' of the fuel.

Thermochemical equations for chemical reactions should be interpreted in molar terms. To compare different fuels, it is useful to write the balanced chemical equations so that only one mole of fuel molecules is involved.

This also allows a simple recipe to be devised for calculating the heat released when one gram of a given fuel is burned.

All of the fuels in Table 6.3, except for hydrogen, are anthropogenic sources of carbon dioxide. The advantages and disadvantages of hydrogen as a fuel are summarized in tabular form in the answer to Question 18. Petrol suitable for modern car engines is obtained via the fractional distillation of crude oil followed by cracking and reforming of the appropriate fractions. The octane rating is an arbitrary scale which reflects the knocking properties of a given petrol; modern high compression engines require ratings of greater than 90. One way of achieving this is to add lead tetraethyl to the refined petrol, but this causes car exhausts to issue airborne lead compounds which damage health. Lead-free petrol is further refined but still uses octane enhancers, e.g. MTBE. Ethanol and methanol can also be used as fuels for motor vehicles.

Photosynthesis in plants combines carbon dioxide and water in the presence of sunlight to give carbohydrates and oxygen. Cellulose (a structural component) and starch (an energy storage system) are both produced. Animals use fats for long-term energy storage. On a mass-for-mass basis, these are far more 'energy-rich' than starch.

Chapter 7
Estimating an enthalpy change

The enthalpy changes quoted in the thermochemical equations in Table 6.3 represent values which were obtained experimentally. If something is measured experimentally, then it is always a challenge to see if it also can be calculated, or estimated, in some way. If this is possible, then it usually turns out that insight can be gained into the factors that determine the value of the measured quantity. For instance, why is it that combustion reactions are so exothermic? In fact, the aim of this Chapter is to try to answer this question; but first we need to see how we can estimate enthalpy changes.

7.1 A hypothetical scheme

Once again we shall take the combustion of methane gas as our example. You can start the discussion of how to make an estimate of an enthalpy change for this reaction for yourself by attempting Question 21.

> **Question 21** You may recall that in Chapter 5 we introduced the idea of an 'enthalpy ladder'. Without looking back, see if you can sketch such a ladder for the reaction corresponding to the complete combustion of methane gas.

When you drew your enthalpy ladder, you simply had to concentrate on the *initial* and *final* states of the reaction. Although this may have seemed quite reasonable to you, it does represent a very fundamental idea and one which we can use to our advantage.

To be more formal, the enthalpy change for a reaction taking place at constant temperature and pressure depends *only* upon the initial and final states of the reactants and the products. This is, in effect, saying that the enthalpy change is *independent* of the way in which the reaction occurs. This is a bit like setting off to travel from one city to another – you can either go via the motorway or via picturesque villages; either way, you finish up the same distance away. Just as there is an element of choice in selecting a route, so we can choose *in a thought experiment* the steps that a reaction takes to reach the final products. These steps need *not* bear any relationship to what actually happens in practice but – and this is the significant point – we can select them so that they are informative about the enthalpy change for the overall reaction. It is important not to confuse in any way the steps we choose to represent a reaction and its true reaction mechanism. The latter, if you like, is reality; what we choose is hypothetical.

A simple, but certainly unrealistic, scheme of events representing the combustion of methane gas is shown in Figure 7.1. The whole sequence takes place in the gas phase. The enthalpy change we wish to estimate is represented by the difference between the product and reactant rungs; it can be represented as ΔH (reaction). (In the discussion that follows, you may find your model kit useful.) The reactants are shown as one molecule of methane and two molecules of oxygen. In **step 1**, these molecules are broken up so that only individual atoms are present – i.e. one carbon atom, four hydrogen atoms and four oxygen atoms. This first step can be represented by the equation

$$CH_4(g) + 2O_2(g) = C(g) + 4H(g) + 4O(g) \qquad \textbf{(step 1)}$$

In **step 2**, there is a regrouping of atoms and one molecule of CO_2 and two molecules of H_2O are formed. This can be represented by the equation

$$C(g) + 4H(g) + 4O(g) = CO_2(g) + 2H_2O(g) \qquad \textbf{(step 2)}$$

Adding the two equations, representing **steps 1** and **2**, together gives the equation for the overall reaction. Of course, the reaction sequence could also be taken to represent the reaction in molar quantities, but to draw all of the molecules involved would be impossible.

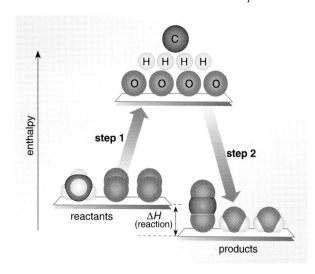

Figure 7.1
A hypothetical scheme, in terms of just two steps, for the combustion of methane gas: $CH_4(g) + 2O_2(g) = CO_2(g) + 2H_2O(g)$. The whole sequence takes place in the gas phase.

▪ If you concentrate just on an oxygen molecule in the first step, can you write down an equation to show what has happened to it?

▪ In the first step, an oxygen molecule is effectively pulled apart to produce two oxygen atoms, i.e. the equation is

$$O_2(g) = O(g) + O(g)$$

▪ If this equation were considered to represent the breaking up or *dissociation* of one mole of oxygen molecules, would you expect the reaction to be exothermic or endothermic?

▪ As you learnt earlier in Chapter 3, an oxygen molecule is at a potential energy minimum with respect to the interactions responsible for chemical bonding. If the individual oxygen atoms are moved apart, then the potential energy increases. Thus, energy will have to be provided to dissociate an oxygen molecule. This means that the reaction is endothermic.

In fact, a thermochemical equation can be written for the dissociation of one mole of oxygen molecules. The enthalpy change measured experimentally is 498 kJ and so

$$O_2(g) = O(g) + O(g) \quad \Delta H = +498\,kJ$$

The enthalpy change in this equation is given a special name, the **bond enthalpy**. In fact, this term is quite general and applies to the dissociation reactions of all molecules which contain just two atoms. Since the dissociation process is always endothermic, the bond enthalpy is always taken to be positive. Table 7.1 provides information on bond enthalpies for a few selected molecules.

Table 7.1 Bond enthalpies for a few selected molecules.

Molecule	Thermochemical equation		Bond enthalpy
oxygen, O_2	$O_2(g) = O(g) + O(g)$	$\Delta H = +498\,kJ$	498 kJ
hydrogen, H_2	$H_2(g) = H(g) + H(g)$	$\Delta H = +436\,kJ$	436 kJ
nitrogen, N_2	$N_2(g) = N(g) + N(g)$	$\Delta H = +945\,kJ$	945 kJ

The idea of a bond enthalpy is useful and, in principle, it can also be applied to molecules that contain more than two atoms. However, a more convenient way of proceeding is to introduce the idea of **average bond enthalpies**. This is described in Box 7.1 overleaf.

▪ Given the information in Tables 7.1 and 7.2, consider if you can now see how to calculate the total enthalpy change for **step 1** in the process shown in Figure 7.1.

▪ The key is to consider the number of bonds that are broken. Thus, there are 4 mol of carbon–hydrogen bonds (each requiring 416 kJ) for methane and 2 mol of oxygen–oxygen bonds (each requiring 498 kJ) for oxygen. Thus, for **step 1**, the total enthalpy change amounts to

$$(4 \times 416\,kJ) + (2 \times 498\,kJ) = 2\,660\,kJ$$

In the calculation that has just been carried out, it is much easier to use a shorthand notation and take 'as read' that the bond enthalpy calculation will be carried out in molar terms *and be based upon the balanced chemical equation exactly as it is written*. Thus, when we consider the bonds broken we could simply write a list:

Bonds broken	*Enthalpy change*
$4 \times (C{-}H)$	$4 \times 416\,kJ = 1\,664\,kJ$
$2 \times (O{=}O)$	$2 \times 498\,kJ = 996\,kJ$
	Total: $2\,660\,kJ$

Note carefully that in this list we have recognized that the chemical bond between the oxygen atoms in an oxygen molecule is a double bond. If necessary, the list can always be checked in a practical way by making models of the reactant molecules and then taking them apart.

Box 7.1 Average bond enthalpies.

The idea of a bond enthalpy in a molecule containing several atoms is that it represents the enthalpy change when a *particular* chemical bond between two linked atoms is broken. Thus, for a water molecule we could imagine breaking just one of the oxygen–hydrogen bonds (Figure 7.2).

By the same token, we could also imagine breaking the oxygen–hydrogen bond in a methanol molecule (Figure 7.3).

In fact, careful measurement shows that the enthalpy changes for these two processes are not the same. If one mole of molecules of each species is considered, then the enthalpy change for the former is +499 kJ, while that for the latter is +428 kJ. Thus, the bond enthalpy depends on the nature of the molecule in which the two linked atoms find themselves.

An average bond enthalpy is simply calculated by taking an average over a large collection of compounds in which a particular linked pair of atoms appears. A selection of average bond enthalpies is given in Table 7.2.

Figure 7.2
Breaking a single oxygen–hydrogen bond in a water molecule.

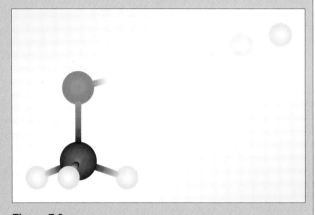

Figure 7.3
Breaking a single oxygen–hydrogen bond in a methanol molecule.

Table 7.2 A selection of average bond enthalpies.

Type of bond	Average bond enthalpy[a]
single carbon–carbon; C—C	330 kJ
double carbon–carbon; C=C	589 kJ
carbon–hydrogen; C—H	416 kJ
oxygen–hydrogen; O—H	463 kJ
single carbon–oxygen; C—O (e.g. as in methanol)	327 kJ
double carbon–oxygen; C=O (e.g. as in carbon dioxide)	804 kJ

[a] These values refer to the enthalpy change for one mole of the polyatomic molecules which contains the bond under consideration.

What can we say about **step 2** in Figure 7.4? Clearly, chemical bonds are no longer broken; in fact, exactly the opposite happens.

▪ Would you expect the formation of a chemical bond between two atoms to be an exothermic or endothermic process?

▪ The overall potential energy will decrease as the two atoms approach one another and form a chemical bond (recall Figure 3.7). The process is thus exothermic.

The dissociation of '1 mol of chemical bonds' has an enthalpy change given by the bond enthalpy. Thus, the reverse process – the formation of '1 mol of chemical bonds' – will have an enthalpy change which is simply the negative value of the bond enthalpy. So, for example if we were to consider the formation of 1 mol of O_2 molecules from O atoms, then the thermochemical equation would be

$$O(g) + O(g) = O_2(g) \qquad \Delta H = -498 \, kJ$$

▪ Using again information from Tables 7.1 and 7.2, can you now calculate the total enthalpy change for **step 2** in Figure 7.1?

▪ If we use the shorthand notation, but notice that we must now consider bonds formed, then the list will be

Bonds formed	*Enthalpy change*
$2 \times (C{=}O)$ (for CO_2)	$2 \times (-804 \, kJ) = -1\,608 \, kJ$
$4 \times (O{-}H)$	$4 \times (-463 \, kJ) = -1\,852 \, kJ$
	Total: $\qquad -3\,460 \, kJ$

Having looked at **steps 1** and **2** in Figure 7.1 in some detail, we are now in a position to add more information to the enthalpy ladder you drew for yourself in Question 21. This is done in Figure 7.4. You should spend a little time looking at this Figure; in particular, check for yourself that it summarizes the calculations that have been carried out.

Figure 7.4
A more detailed enthalpy ladder for a hypothetical scheme representing the combustion of methane gas. It shows schematically the breaking up of the reactant molecules into atoms and then a regrouping of these atoms to form product molecules.

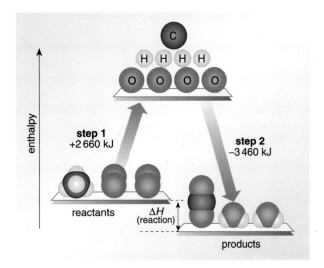

▨   What happens if the chemical equations representing **steps 1** and **2** (p. 246) are added together?

■   The addition gives

$$CH_4(g) + 2O_2(g) + C(g) + 4H(g) + 4O(g) = C(g) + 4H(g) + 4O(g) + CO_2(g) + 2H_2O(g)$$

Cancelling out the species that are the same on both sides of the equation gives the chemical equation for the overall reaction, i.e.

$$CH_4(g) + 2O_2(g) = CO_2(g) + 2H_2O(g)$$

Just as the two chemical equations that represent **steps 1** and **2** can be added together to give the chemical equation for the overall reaction, then so can the enthalpy changes be treated in a similar manner. Thus, the overall enthalpy change is given by the sum of the individual enthalpy changes for the two steps,

$$\Delta H \text{(reaction)} = \Delta H (\textbf{step 1}) + \Delta H (\textbf{step 2})$$

This relationship can also be seen from the way in which Figure 7.4 is drawn.

▨   What is the overall enthalpy change for the complete combustion of 1 mol of CH_4 molecules according to the information summarized in Figure 7.4?

■   The overall enthalpy change is given by

$$\Delta H \text{(reaction)} = 2\,660 \text{ kJ} + (-3\,460 \text{ kJ})$$
$$= 2\,660 \text{ kJ} - 3\,460 \text{ kJ}$$
$$= -800 \text{ kJ}$$

As expected, the overall reaction is exothermic. Because the calculation has involved *average* bond enthalpies then the answer can only be considered to be an estimate; nonetheless, the agreement with the value given in Table 6.3, i.e. −804 kJ, is very good.

The approach that has been used to estimate the enthalpy of combustion of methane gas is quite general and the simple two-step scheme shown in Figure 7.1 can be adapted for any other fuel. The estimate of the enthalpy of combustion will be based, however, on both the fuel and the water formed *being in the gaseous state*. This may not be the case in practice and, in such circumstances, corrections to the estimate are needed; but usually, and certainly for our purposes, these can be taken to be small.

Does the approach we have used to estimate an enthalpy of combustion provide any insight into why combustion reactions are exothermic? You may recall this was a question that was asked at the start of this Chapter.

▨   One major clue lies in the list of average bond enthalpies in Table 7.2. Can you suggest what this is?

■   The average bond enthalpy for the carbon–oxygen bond, as found in carbon dioxide, stands out as being very large.

Thus, to break a carbon–oxygen bond in carbon dioxide – an endothermic process – requires a lot of energy but, by the same taken, to form such a bond – an exothermic process – releases a lot of energy. If we think about the combustion of a hydrocarbon purely in terms of bond making and breaking, then the formation of carbon dioxide is going to be a key factor in accounting for the exothermic nature of the reaction. Looking at Table 7.2 in a little more detail shows that the oxygen–hydrogen bond as found in water – the other product of combustion – also has a relatively high bond enthalpy compared to carbon–carbon and carbon–hydrogen bonds. Again this will favour an exothermic combustion reaction for a hydrocarbon. In summary, although we may think of a fuel as a 'store of energy', it is the reaction with oxygen to yield carbon dioxide and water that is so vital in accounting for the exothermicity. We have discovered a key reason that explains why combustion reactions are so exothermic.

Summary of Chapter 7

This Chapter has been concerned with the procedures for estimating an enthalpy change. It has introduced the ideas of bond enthalpy and average bond enthalpy. The best way to summarize the Chapter is to work through a further example. You are asked to do this for yourself in Activity 9.

Activity 9 Estimating an enthalpy change

Using the approach described in this Chapter, make an estimate of the enthalpy change for the complete combustion of octane gas according to the chemical equation

$C_8H_{18}(g) + 12\tfrac{1}{2}O_2(g) = 8CO_2(g) + 9H_2O(g)$

(In the octane molecule (Figure 7.5), there are 18 carbon–hydrogen bonds, and do not forget the carbon–carbon bonds.) How does your result compare with that in Table 6.3?

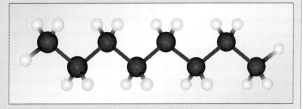

Figure 7.5
The octane molecule.

Chapter 8
The direction of chemical change

In previous Chapters, we have been primarily concerned with practical aspects of chemical reactions and so have not been concerned with the fact that some chemical substances will not actually react with one another. For example, a mixture of carbon dioxide gas and water vapour will not produce methane gas and oxygen gas, and nor will a glass of water of its own accord decompose into hydrogen gas and oxygen gas. We might view this as 'just one of those things' but to do so would be to fail to ask what turns out to be a very basic question: 'Why do some chemical reactions occur and yet others do not?'.

Perhaps surprisingly, the answer to the question can be framed in terms of very fundamental physical, rather than chemical, processes. To develop the arguments in detail would be complex but it is worthwhile digressing a little to provide at least a flavour of the answer. The approach we shall take is very much based on that taken in the book *Atoms, Electrons and Change* by Peter Atkins.

To set the scene, it is useful to return to an example of a chemical reaction we introduced in Chapter 2. Here, as you may recall , we saw that the chemical reaction between shreds of aluminium and liquid bromine was very vigorous (Figure 2.2). The reaction started as soon as the two substances came into contact and could be described as being **spontaneous**. Would you say that the reaction between methane gas and oxygen gas was also spontaneous?

In all probability you would have said no. In terms of interpreting the word 'spontaneous' in an everyday way, you are undoubtedly correct. But, unfortunately, chemistry is a little perverse in this area. The word 'spontaneous' is given a deeper meaning and is used to describe any reaction that has '*a natural tendency to occur*'. The reaction between methane gas and oxygen gas (to give carbon dioxide and water) does possess this natural tendency. The fact that it requires the stimulus of a spark or flame is not relevant to our view of its spontaneous nature. We know that the reaction is poised for change and on this basis describe it as spontaneous. To take another example, you might be tempted to argue that the decomposition of water into oxygen gas and hydrogen gas is a spontaneous reaction because it can be made 'to go' in an electrolysis experiment. But turn off the current and the reaction stops; there is no *natural* tendency and therefore it is not spontaneous. The term spontaneous must also not be confused in any way with the speed of a reaction. A chemical reaction that occurs very slowly, e.g. the rusting of iron, is just as spontaneous as one that occurs explosively fast.

To summarize, the term 'spontaneous' applied to a chemical reaction is used in the sense that the reaction has a natural tendency to occur; if you like, it suggests that there is an inherent property that favours change. The description takes no account of how the reaction is started nor how fast it proceeds.

Box 8.1 Nitroglycerine – the first high explosive

Figure 8.1
The manufacture of nitroglycerine in the 1890s. The drawing is of the 'hill-man' (see text) carrying out the nitration process; notice the one-legged stool.

Ascanio Sobrero, an Italian chemist, discovered nitroglycerine in 1846; it was the first high explosive and signalled a technological revolution. The material itself is a heavy oily liquid with molecular formula $C_3H_5N_3O_9$ (Figure 8.2).

It can be distilled or solidified in reasonable safety. However, detonation can readily be brought about by a sudden, but relatively small, energy input, e.g. a sudden impact or rapid local heating (a hot spot). This particular explosive character of the material was harnessed by the inventiveness of the Swedish chemist and businessman, Alfred Nobel. He devised a 'detonator' and solved the problem of handling the liquid by absorbing it on 'diatomaceous earth' to give the familiar 'sticks of dynamite'. Modern explosives for blasting still use nitroglycerine (Figure 8.3).

The early manufacturing methods of nitroglycerine were uncomplicated. The liquid was produced by treating purified glycerine with an acid mixture - a procedure called nitration. Following this process, transfer of nitroglycerine to the rest of the production plant was then by gravity-feed and so the 'nitrator' had to be located on a natural, or artificial, hill. The workers involved in this delicate part of the process thus came to be known as 'hill-men'. To ensure that they remained alert at their posts, they used one-legged stools.

The decomposition of nitroglycerine, even though it needs a stimulus, is a spontaneous reaction. The essential reaction between the 'fuel' and oxygen is conducted *within* the structure of the molecule itself. In other words, the oxygen-rich structure of nitroglycerine supplies its own oxygen for combustion. The decomposition once started takes place very rapidly and gives rise to a whole host of small gas molecules; in fact, there is a sudden 1200-fold expansion in volume and this produces a pressure wave that provides the destructive shock of a 'high explosive'.

$$4C_3H_5N_3O_9(l) = 6N_2(g) + O_2(g) + 12CO_2(g) + 10H_2O(g)$$

Figure 8.2
The nitroglycerine molecule.

Figure 8.3
Sticks of high-density, nitroglycerine-based explosive used in blasting rock, for example.

An equivalent question to asking, 'Why do some chemical reactions occur while others do not?' is simply to ask 'Why are some chemical reactions spontaneous?'. One suggestion that might have occurred to you is to argue by analogy with mechanical situations. Thus, if we place a ball on a slope we know that it has a 'natural tendency' to move downhill, that is in the direction which leads to a decrease in potential energy. So, we might argue that a spontaneous chemical reaction is also one that leads to a decrease in energy. It is certainly true that many spontaneous reactions are exothermic and so they release energy to their surroundings. However, there is a problem; you may have realized what it is already. Despite the fact that exothermic reactions are in the majority, we cannot simply 'sweep under the carpet' those reactions that turn out to be endothermic; after all, they are chemical reactions just the same. In terms of a mechanical analogy, endothermic reactions would correspond to a ball on a slope having a natural tendency to roll uphill! We have to accept that, even though the idea may at first sight be appealing, there is something more fundamental about the spontaneity of a chemical reaction than energy considerations alone can reveal.

Two extracts from Peter Atkins' book indicate how to begin to deal with spontaneity:

> *The tendency of a reaction to form products is governed by that most remarkable principle of nature, the 'Second Law' of thermodynamics. The Second Law was established by Rudolph Clausius in 1850 and by William Thomson, Lord Kelvin, in 1851, but its applications to chemical reactions were not fully recognized until the work of Josiah Gibbs in the 1870s came to be appreciated at the turn of the century.*
>
> *The Second Law is concerned with 'spontaneous changes', the changes that have a natural tendency to occur …*

and, in a little more detail,

> *… the Second Law is, in brief, a summary of the tendency of matter and energy to disperse in a more disorderly form. That is, the law recognizes that nature has a natural tendency to lose whatever orderliness it currently possesses and to collapse into greater disorder: the natural tendency of change is toward decay.*

The subject of thermodynamics, which plays an important role in many areas of science and engineering, is concerned with 'transformations of energy'. It is a complex, inherently mathematical, subject; some find it fascinating, but many do not! The fact that there is a 'Second Law' implies there must be a 'First'. Although we did not state it explicitly, a popular statement of the latter, 'Energy is conserved' was introduced in Chapter 3.

What is meant by 'matter and energy dispersing in a more disorderly form'? To deal with the dispersal of matter, we can imagine injecting a small amount of gas into one corner of a sealed container (Figure 8.4). We know that in time, the chaotic motions of the molecules within the gas will ensure that it spreads throughout the whole container. The likelihood that the gas molecules will ever all collect again into the

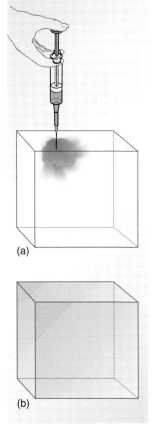

(a)

(b)

Figure 8.4
A gas injected into a sealed container (a) will eventually spread (b) throughout the whole container. There is a spontaneous physical change involving the dispersal of matter.

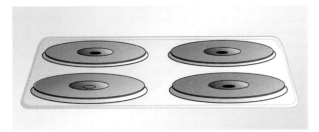

Figure 8.5
A hot block of metal will cool down. There is a spontaneous physical change involving the dispersal of energy.

corner in which they were injected, that is in a sense 're-order', is negligible. So, in the container, because of the chaotic motions of the gas molecules, there is a natural tendency towards the dispersal of matter.

The dispersal of energy as a natural physical process can be demonstrated by considering a hot block of metal standing in air (Figure 8.5). We know that the block will cool down. At the molecular level in the hot block (as described in Book 1), the atoms will be undergoing a 'frantic, getting nowhere motion'. Incessantly, atoms will jostle other atoms and, in particular, at the metal surface molecules in the air will be jostled and fly off never to return. In this chaotic fashion, energy is dispersed as heat into the surroundings. The chance that molecules in the air will be able to 're-order' and jostle atoms in the cooling block so that it can accumulate energy, and hence warm up again, is negligible. So, for the hot block of metal there is a natural tendency towards the dispersal of energy.

The tendency of matter to disperse in a disorderly fashion and the tendency of energy to do likewise can both be regarded as fundamental spontaneous processes. In fact, Atkins uses the term 'primitive spontaneities' to describe them.

The Second Law itself works in the currency of 'entropy'. This concept allows the spontaneity of a process to be discussed on firm quantitative foundations. For our purpose, we shall simply take it as a measure of the dispersal of matter and energy. In effect, entropy is taken to be synonymous with disorder; an increase in disorder is accompanied by an increase in entropy. In these terms, we can begin to understand the often-quoted statement due to Clausius that '*The entropy of the Universe tends to increase*'. It implies that the Universe is continuously sliding into disorder. (In philosophical terms, the implication is that all of the energy and matter in the Universe will eventually be dispersed and no further changes will occur!)

If the natural tendency is to disorder, then why is it, say, that structurally complex molecules can be made from simpler parts – either in our bodies or at the laboratory bench? The answer, at least in part, relates to the way the term 'Universe' in Clausius' statement must be interpreted. It literally means 'everything', that is in terms we have already used in Chapter 5, the system *and* the surroundings. It is perfectly acceptable for there to be an increase in *local* order but a careful analysis of all of the processes that are involved in some way with the reaction in question will reveal that the total entropy for system and surroundings does increase. At this point, the discussion starts to enter difficult territory and we shall not take it further. Rather we shall look at how Atkins treats two well-defined chemical examples.

The first is one with which we are already familiar; the combustion of methane in oxygen:

$$CH_4(g) + 2O_2(g) = CO_2(g) + 2H_2O(g)$$

... why is the combustion reaction spontaneous, but not its reverse? There is not a great deal of difference in the complexity of the reactants and the products of the methane combustion reaction, since three small gaseous molecules are replaced by three other small gaseous molecules. Therefore, if we improperly confined our attention to the disorder of the matter, we would conclude that in the combustion reaction there was little change in the disorder of the Universe as it proceeded, and hence little reason why the combustion should occur at all. (Detailed calculation shows that there is in fact a small decrease in disorder.)

How do you think that the discussion continues? It focuses on the fact that the combustion reaction releases a great deal of energy as heat into the surroundings. This amounts to a very considerable dispersal of energy. The fact that energy is released is important but it is the dispersal that accounts for the spontaneity.

Hence, overall the Universe is more chaotic after methane has burned than it was before, and so the combustion reaction is spontaneous.

The second example that is discussed is the dissolution of ammonium nitrate NH_4NO_3 in water:

$$NH_4NO_3(s) = NH_4^+(aq) + NO_3^-(aq)$$

It is an endothermic process; but this is not out of the ordinary. The dissolution of any ionic solid can be viewed in terms of two *hypothetical* steps as shown in Figure 8.6. In the first step, the solid is broken down into individual ions which enter the gas phase – ionic interactions have to be overcome and so this step requires energy. In the second step, the individual ions in the gas phase interact with liquid water and enter into solution – a process which releases energy. The Figure, though hypothetical, is realistic in that it depicts a relatively fine balance between the enthalpy changes for the two steps. For ammonium nitrate, it turns out that the balance results in an endothermic process; for a different ionic compound, the dissolution process could well be exothermic.

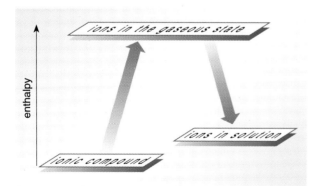

Figure 8.6
Dissolving an ionic solid divided into two hypothetical steps. The overall dissolution process will be endothermic in this case.

So, why is the dissolution of ammonium nitrate a spontaneous process when it absorbs, rather than disperses, energy? We must also consider the dispersal of matter. When ammonium nitrate dissolves in water, the ions not only disperse in solution but they also break down the orderly hydrogen-bonded arrangement of water itself. It is this *overall increase* in the disorder in solution that outweighs the *decrease* in disorder due to energy being absorbed from the surroundings and, consequently, the dissolution is spontaneous.

> *The spontaneity of an endothermic reaction is not perplexing once we acknowledge that the direction of spontaneous change is determined by entropy, not energy, and in particular by the overall disorder of the Universe.*

So, this Chapter has considered the very basic question, 'Why do some chemical reactions occur and yet others do not?'. Because of the complexity, no more than a flavour of the answer has been given. (It is for this reason that specific questions have not been asked.) It is hoped, however, that sufficient has been said to enable you to appreciate a very fundamental way of thinking about chemical reactions although the discussion has not gone far enough to allow you to make predictions concerning spontaneity for yourself.

Summary of Chapter 8

A spontaneous chemical reaction is one that has a natural tendency to occur. The term spontaneous takes no account of how a reaction has to be initiated nor how fast it proceeds. The decomposition of nitroglycerine is a spontaneous reaction that requires a sudden, but relatively small, energy input to initiate it. Nitroglycerine is an example of one of the first 'high explosives'.

The tendency of a reaction to form products is governed by the Second Law of thermodynamics. In brief, this summarizes the tendency of both matter and energy to disperse in a disorderly fashion. The tendency of matter to disperse in a disorderly fashion and the tendency of energy to do likewise can both be regarded as fundamental spontaneous processes. The combustion of a fuel, or the dissolution of a simple ionic compound in water, can both be discussed in terms of the fundamental processes of energy and matter dispersal. The concept of entropy provides a firm foundation for the Second Law. For a spontaneous change, the total entropy for system and surroundings (i.e. the Universe) will always tend to increase.

Chapter 9
Energy in the environment

9.1 A brief survey of the Earth's energy sources

The landscape (Figure 9.1) summarizes the Earth's energy sources; most of these in one way or another depend on the radiation provided by the Sun. Some of the sources have been discussed already. Thus, sunlight is essential, via the process of photosynthesis, to the natural growth of plants and these, in turn, lie at the heart of the intricate webs of the food chain. The fossil fuels, on which so much of the energy needs of developed countries currently depend, are essentially stores of the Sun's energy from long ago. But we should not forget that biomass, in particular fuelwood, remains a principal source of energy in less developed countries; perhaps accounting for as much as 10–15% of total world energy use.

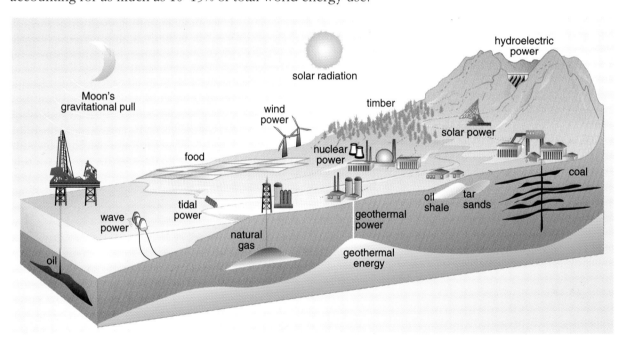

Indirectly, it is also the Sun's energy that is responsible for the restless energy of the Earth, that is the waves and currents of the oceans, the patterns of the wind and the flow of the rivers. Just over 20% of the Sun's energy that reaches Earth goes into driving the vast hydrological cycle. The technology for taking advantage of the energy of the wind, or of fast-flowing or falling water, may be new, but the ideas are not. The Domesday Book recorded no fewer than 5 624 watermills in Britain. The earliest windmills are reputed to have been in use in China over 2 000 years ago and sail windmills were widespread in 19th century Britain (Figure 9.2). Modern technology is also making it possible, but presently on a small scale, to tap the Sun's energy directly via solar batteries and thermo-electric power stations.

Figure 9.1
The main energy sources of the Earth.

Figure 9.2
In the 19th century, Britain had roughly 10 000 sail windmills in operation; a large one could generate up to 30 kW which was sufficient for the milling needs of a small community. Recently, there has been a resurgence of interest in wind power, with most modern windmills being of the 'propeller type'; they are usually referred to as wind turbines.

Our dependence on the Sun for energy is immense and varied; we no longer worship it as in many early religions but its influence is all pervading. But more than this, the manner in which the Sun's radiation *interacts* with the Earth's atmosphere is vital to life itself; it provides the climate in which plant, animal and human life thrives.

There are non-solar sources of energy. Of these, the major source is nuclear energy whose origins can be traced back to cosmic processes preceding the formation of the Solar System. Two other sources are tidal energy arising, primarily, from gravitational effects of the Moon (with some contribution from the Sun), and geothermal energy which flows from the heat trapped deep inside the Earth itself. It is estimated, for instance, that the total energy available in the UK from the latter source could be comparable to that stored in the nation's coal reserves. Finally, there is one other tantalizing source of energy to mention. This is nuclear fusion which is also the source of the Sun's energy. In the briefest of terms, nuclear fusion is the combination of very light nuclei – such as the isotopes of hydrogen, deuterium and tritium, to produce helium – with the result that overall there is a decrease in mass which manifests itself as a release of energy: very large amounts of energy, indeed. If this process could be controlled on Earth in a 'fusion reactor', then the benefits would be enormous. Currently, there are national, and international, projects to build such reactors but the technical problems are formidable (Figure 9.3).

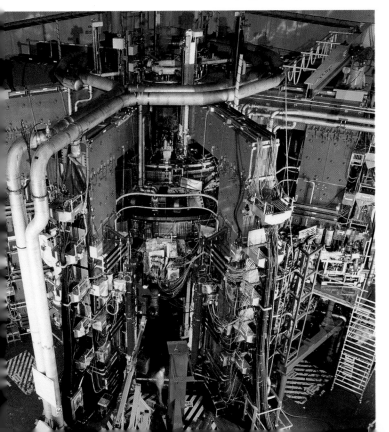

In practice, a distinction is often made between **'renewable'** and **'non-renewable' energy resources**. To define the latter is more clear cut; they are the finite resources of fossil fuels and, in addition, the materials – such as uranium ores – of the nuclear industries. These are the resources that will be exhausted at some point in the future, because nature cannot produce more of them, at least on a human time-scale. Under the umbrella of renewable – or non-depleting – sources comes 'the Sun, the rivers, the tides and waves of the sea, the wind, the Earth's internal heat and the natural growth of plant life'. But the term is a little 'grey' for it is strictly only the Sun, the wind and waves which will continue for generations to come.

Figure 9.3
The Joint European Torus (JET) sited near Abingdon, Oxon, UK. JET represents the culmination of intense research effort over the years and is a major step towards exploiting controlled nuclear fusion as an energy source for the future.

Although brief, this portrait of the Earth's energy sources should help you to appreciate that although the canvas is of global proportions, very basic chemical and physical processes are at work: you have come across various examples of these already in this part of the Book. However, new ideas emerge when we take a closer look at the way in which radiation from the Sun interacts with the Earth's atmosphere. It is these ideas we focus on in this concluding Chapter. As a starting point, it is important to have a better understanding of what we mean by the term 'radiation'.

9.2 Electromagnetic radiation

Although we may not call it electromagnetic radiation, we are familiar with various forms of it – sunlight, torch beams, the X-rays used for diagnosis in a hospital, the microwaves in our microwave ovens, and the radio waves that allow us to listen to our favourite radio programmes. These beams, rays and waves – as we call them – can all be described under the single heading of **electromagnetic radiation**. This generic term applies to all forms of radiation including X-rays and gamma-rays (γ-rays), ultraviolet, visible light (the light we can see), infrared, microwave radiation and radio frequency radiation. Of these, it is ultraviolet, visible light and infrared that will be our main concern in this Chapter.

A common property of all electromagnetic radiation, although the types may seem to have very different characteristics, is that it passes through space at a constant speed; this is the speed of light, 3.0×10^8 m s^{-1} in a vacuum. The term electromagnetic is used because the radiation consists of a wave of oscillating electric and magnetic components. However, for our purposes we only need focus on certain features of the wavelike character. In fact, simply looking at water waves will provide the concepts we need.

Throwing a pebble into a still pond of water produces ripples – or waves – which spread out from where the pebble hit the surface. An article floating on the pond, however, does not move along with the waves, it simply bobs up and down. Figure 9.4a shows the general shape of a water wave at two different times. The distance between successive crests (or troughs for that matter) at any given time is called the **wavelength** and it doesn't change with time. Wavelength is given the symbol λ (Greek letter lambda). Compared to the fixed post, the wave appears to move forward, but the water itself simply moves up and down – this is why the article in the pond just bobs up and down. Figure 9.4b demonstrates the water moving up and down by plotting the height of the water at the post with respect to the still water level at successive intervals of time.

Frequency is a measure of the number of wavelengths that occur in a fixed period of 1 s; it can be given the symbol f. The unit of frequency is 'hertz' (symbol Hz) or 'per second' (symbol s^{-1}). Thus, a frequency of 10 Hz is interpreted as something that happens ten times in one second. In the case of radiation, the 'something' that happens corresponds to a wavelength.

For all waves, there is a simple relationship between wavelength and frequency. For electromagnetic radiation, it is

$$\lambda = \frac{c}{f}$$

where the symbol c represents the speed of light.

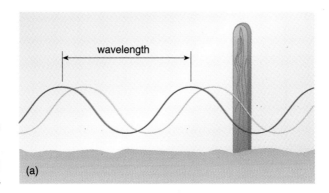

Figure 9.4
(a) The general shape of a water wave at two different times. The post is fixed in the water. The wavelength is marked for the wave in its first position. (b) The height of the water at the post with repect to the still water level at successive time intervals.

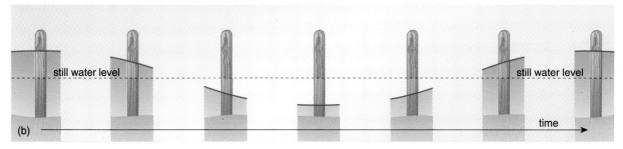

■ Can you see what happens in this equation if the frequency f is increased?

■ The important point to remember is that the speed of light is a constant. Thus, if f increases, the term on the right of the equation must become smaller; i.e. the wavelength gets shorter.

To summarize,

The higher is the frequency of the radiation, the shorter the wavelength.

The regions of the electromagnetic spectrum are not sharply defined, but the general areas are summarized in Table 9.1. In this table, wavelengths are either expressed in metres, or the much smaller unit of nanometres.

Table 9.1 The regions of the electromagnetic spectrum[a].

Region	Wavelength
radio frequency	> 0.3 m
microwave	0.3–0.003 m
infrared	0.003 m–$1\,000$ nm
visible	700–400 nm
ultraviolet	300–3 nm
X-rays and gamma-rays	< 3 nm

[a] The infrared and ultraviolet regions are often divided into 'near' and 'far' areas where 'near' refers to wavelengths close to the visible region. There are no 'gaps' in the spectrum: it is just typical wavelength regions which are shown.

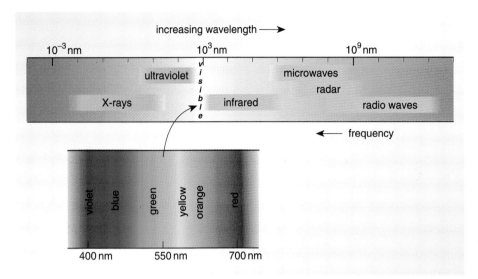

Figure 9.5
A portion of the electromagnetic spectrum. The wavelengths indicated in the expanded visible part of the spectrum are typical, but in approximate positions. Note that they also correspond to different colours. It's worth remembering that ultraviolet radiation is radiation of shorter wavelength (higher frequency) than violet light ('ultra' is Latin for 'beyond'). Infrared radiation is radiation of longer wavelength (lower frequency) than red light ('infra' is Latin for 'below').

A portion of the electromagnetic spectrum in the infrared–visible–ultraviolet region is shown in Figure 9.5. The relationship between colour and wavelength is emphasized by the expansion of the visible part of the spectrum. This part is so-called because it corresponds to the range of electromagnetic radiation which our eyes can detect; roughly from 700 nm (red light) to 400 nm (violet light). 'White light' is a mixture of all of the visible components, as will be discussed in greater detail in Book 4.

An important feature of electromagnetic radiation is that it 'carries energy'; indeed, the term 'radiant energy' is commonly used. In terms of a physical picture, we understand this by representing the radiation as a stream of particles called 'photons'. According to this model, each photon is taken to represent a packet of energy. (This is the important point for our discussion and we shall not enter into the complex area that considers the relationship between the particle and wave descriptions of radiation.) In more everyday terms, our own experience tells us that radiation carries energy: we can feel the warming effect of radiation from infrared lamps, or even from our desk lamp, and we can cook with microwaves.

It is a well-established principle that the photons associated with higher frequency radiation have more energy associated with them than those of lower frequency radiation, e.g. high-frequency X-rays do more damage than radio waves. If we restate this in terms of wavelength, then for a given photon the shorter the wavelength, the higher the energy. The intensity of a beam of radiation, and hence how much energy it carries, depends upon the number of photons it contains.

Question 22 Which has the higher energy, blue light or red light, if both have the same intensity?

9.3 Solar radiation and the Earth

9.3.1 The greenhouse effect

The Sun provides the Earth with a continuous supply of energy which it is estimated will last for a further 4 to 5 billion years; this is despite the fact that it is losing mass at the unimaginable rate of 6 million tonnes per second. The energy, which is in the form of electromagnetic radiation, arises because even though within the Sun the temperatures reach well above one million degrees Celsius, the surface temperature is around 6 000 °C. As described in Box 9.1, all hot objects emit radiation depending on their surface temperature.

Box 9.1 Hot objects emit radiation

The idea that something that is hot emits radiation is supported by common experience. A bar of iron (Figure 9.6), heated to a temperature at which it 'glows', cools down by a number of mechanisms, one of which includes the emission of radiation. This radiation is primarily in the infrared region – we can 'feel' the heat given off; but also there is a proportion in the visible region – we can see this radiation. These observations also tell us that the radiation given off by a hot object is not confined just to an emission in a single part of the electromagnetic spectrum: it is spread over a range of wavelengths. In fact, it is possible to model what happens. If we assume that a 'body' is capable of absorbing and emitting all wavelengths of radiation with equal effect – for convenience we refer to it as an **ideal emitter** – then the manner in which it emits radiation when it is at three different temperatures is shown in Figure 9.7.

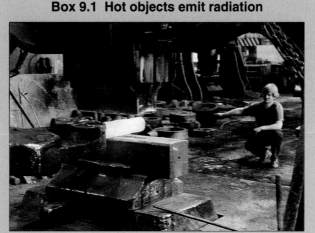

Figure 9.6
Working with very hot iron.

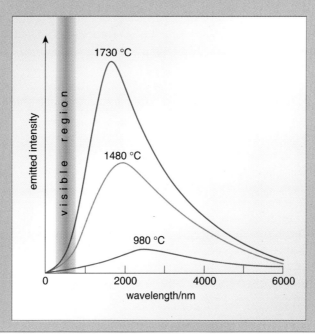

If you look carefully at the Figure, you will see that there is a tendency for the emitted radiation to 'peak' at shorter wavelengths as the temperature is raised. Is this consistent with common experience? If we heat a red hot iron bar further, what happens?

It will become almost 'white hot' if heated sufficiently strongly. In other words, the emitted visible radiation is now not just from the red end of the spectrum but contains a 'mix' of all of the colours and, hence, appears to be 'white'. Evidently, there is an increase in the proportion of shorter wavelength visible radiation (towards the blue) that is emitted as the bar becomes hotter – and this is just what would be predicted from Figure 9.7.

Figure 9.7
The intensity of the emitted radiation (measured in arbitrary units) calculated at three different temperatures for a body that is capable of absorbing and emitting all wavelengths of radiation with equal effect; i.e. an ideal emitter.

The Sun is not a perfect 'ideal emitter' but it is still reasonable to model it as such with a temperature equal to its surface temperature. The form of the emitted radiation for this model is shown in Figure 9.8.

■   How would you describe the emitted radiation?

■   The radiation peaks – or is at its most intense – near the visible region of the spectrum. However, there are still significant amounts of radiation in both the infrared and in the near-ultraviolet regions.

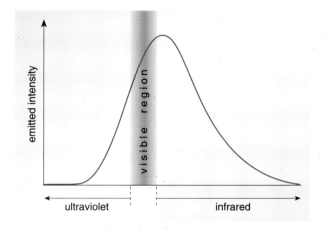

Figure 9.8
A calculation of the radiation emitted from the Sun assuming it to be an ideal emitter with a surface temperature of close to 6 000 °C. (In practice, there is much greater emission in the X-ray, far-ultraviolet and radio-frequency regions than this simple model predicts.)

At the edge of the Earth's atmosphere, the intensity of the incoming solar radiation is close to 1.4 kW per square metre. A sizeable proportion of this radiation – of the order of 30% – is reflected directly back into space by the atmosphere and the Earth's surface; typical reflectors being clouds, atmospheric dust, the molecules of the atmospheric gases, snow, and land not covered by vegetation. The remaining radiation is absorbed and is responsible for maintaining an average global surface temperature of about 15 °C; in crude terms, it is the circulation of the atmosphere that distributes the 'warming effect' around the globe.

■   If the average temperature remains effectively constant, what conclusion can be drawn? (Remember that the Earth will emit radiation itself.)

■   The constant average temperature implies that there must be a balance between the incoming solar radiation which is absorbed and the radiation which is emitted from the warm Earth – i.e. the land, the ocean surfaces and the atmosphere.

The idea of a balance is important. But more critical is an understanding of how it is achieved. (It is cause for concern that humans may have interfered with this balance too much already.)

To begin to understand the balance, it is useful, as with other scientific problems, to develop a model: the simpler the better, to begin with. As a start, we can assume that the Earth's atmosphere does not absorb any radiation. In this picture, we have to find the effective temperature that the land, low-lying atmosphere and ocean surfaces must reach in order for there to be a dynamic balance between incoming and outgoing radiation. Basically, although we shall not do the calculation, a hemisphere of the Earth is taken to be bathed in (non-reflected) sunlight and, in turn, the Earth – taken to be an ideal emitter – emits radiation in *all* directions back into space. The calculated effective temperature is a surprise; it turns out to be close to −18 °C. This is some 33 degrees Celsius below the actual mean temperature and certainly would not be sufficient to support life as we know it. Clearly, there is something missing from the model. But what?

The missing feature is the manner in which certain gases in the atmosphere interact with radiation. In other words, the assumption that the Earth's atmosphere does not absorb radiation is flawed. As a matter of information, it is also important at this stage to recognize that the spread of wavelengths emitted from the Earth's surface – and close to it – is *entirely* in the infrared region. (This is a reflection of the fact that the Earth has a relatively low surface temperature (Figure 9.9).)

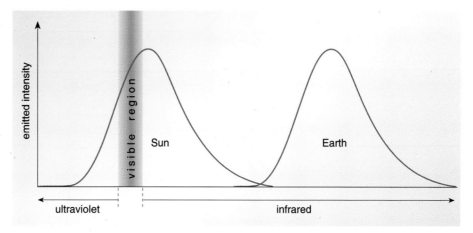

Figure 9.9
A schematic view of the form of the emitted radiation as function of wavelength for the Earth. For comparative purposes, a similar plot for the Sun is shown.

The main constituents of the air are nitrogen N_2 and oxygen O_2, which account for roughly 70% and 21% of the overall composition by volume respectively. (These figures vary slightly because of the variable water vapour content of the atmosphere.) Neither of these gases absorbs either visible or infrared radiation – they are said to be *completely transparent* to these types of radiation. And now the crucial point. Within the remaining constituents, there are several gases which are transparent to visible radiation but *not* infrared radiation. These gases, or strictly their molecules, can interact with and absorb this latter type of radiation (see Box 9.2, overleaf).

So, how do we develop further our model to account for the habitable temperature of the Earth's biosphere? We must now imagine that the atmosphere allows the shorter-wavelength radiation from the Sun, which is not immediately reflected, to pass through unimpeded to be absorbed by the Earth's surface. This surface becomes, as a consequence, warmer and in turn emits radiation. But what type of radiation? And what will happen to this radiation?

If you look back at Figure 9.9, you will see that the emitted radiation from the Earth's surface is all in the infrared region (long wavelength). We can expect that a proportion of this radiation will be absorbed by gases in the atmosphere and stored in their molecular vibrations. This process of 'radiation trapping' can happen a number of times. The gases that absorb radiation also re-radiate it – over the same range of wavelengths – and there is further absorption: either by the Earth's surface or once more by gases in the atmosphere. The gases that are involved are called '**greenhouse gases**' and the overall process – which is complex – is shown schematically in Figure 9.12 (overleaf). The net effect (and we shall not dwell on the details) is to cause an *increase* in temperature at ground level of the order of 33 degrees Celsius above that which it would otherwise be if the greenhouse gases were absent.

Box 9.2 Molecules and infrared radiation

Molecules are not rigid, as the framework models that are used to represent them might imply. For example, in the hydrogen chloride molecule (Figure 9.10) the chemical bond between the hydrogen and chlorine atoms can be quite realistically represented by a spring. This spring, although of molecular proportions, will stretch and contract just like any other spring. Thus, internal motion is possible within the hydrogen chloride molecule and we refer to this as a vibration.

The molecular spring in the HCl molecule has two very important properties. First, it can only vibrate at a specific frequency. Secondly, this natural frequency corresponds with that of radiation in the infrared region. As a consequence the molecule can interact with, and absorb energy from, infrared radiation; but the frequency of this must match that of the spring exactly. In this way, the molecule absorbs energy and it is stored in the vibrational motion. As you might realize, in a sample of hydrogen chloride gas the energy stored in the vibrational motions of the individual molecules will contribute to the total internal energy of the sample.

Most molecules contain more than just two atoms and, consequently, there will be a number of different ways in which a given molecule can vibrate. In addition, the springs representing different types of chemical bond will have differing degrees of 'stiffness'. The vibrations will be now more complex and a typical type of vibration – stretching, bending, rocking or twisting – will involve several atoms all moving

in harmony rather than the stretching of just individual single bonds. The different ways in which a molecule can vibrate will mean that different frequencies of infrared radiation will be absorbed. In fact, this is an important analytical property. The infrared frequencies absorbed by a molecule are very characteristic so that a plot of them as a function of frequency – an **infrared absorption spectrum** – can be regarded as a

'fingerprint' for identification purposes. That is, different substances exhibit different 'infrared signatures'. This is the basis of the technique known as *infrared spectroscopy*.

An example of an infrared spectrum is shown in Figure 9.11. The spectrum is that of dichlorodifluoromethane, CCl_2F_2, which is alternatively known as CFC-12 or Freon-12.

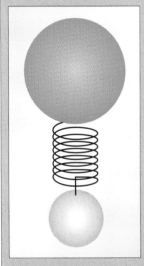

Figure 9.10
A representation of a hydrogen chloride (HCl) molecule (as found in the gaseous state) in which the single chemical bond is represented by a simple spring.

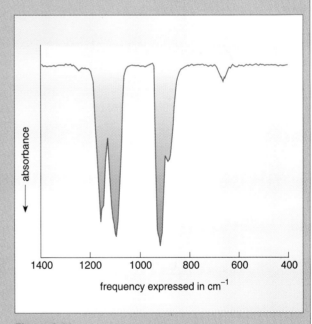

Figure 9.11
The infrared absorption spectrum of CCl_2F_2. Note that the absorption 'peaks' in an infrared spectrum point downwards. Rather than giving the horizontal scale directly in frequency units, it is the convention to divide the frequency of absorption by the speed of light (expressed in cm s^{-1}). This gives relatively small numbers with units of cm^{-1}.

Figure 9.12
A simplified diagram illustrating the greenhouse effect. The reasons for the Earth having a habitable biosphere have been recognized for many years. For example, a paper written in 1857 by Irish-born scientist John Tyndall makes the comment, 'the atmosphere admits the entrance of solar heat, but checks its exit; and the result is a tendency to accumulate heat at the surface of the planet'.

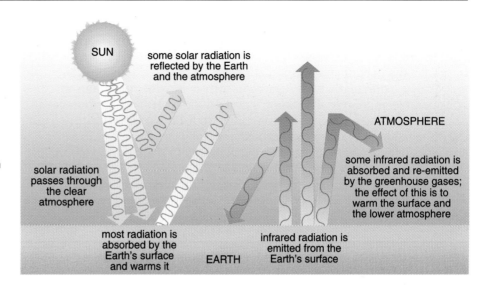

The so-called 'greenhouse gases' result in the **greenhouse effect**, and the effect is a *natural* one (Figure 9.12). It cannot be emphasized too strongly that the effect is essential to life; sometimes, the way in which it is written about does not make this clear. But there is a potential problem: humans are probably interfering with the process in an uncontrolled and not fully understood manner. The term 'greenhouse effect' is slightly misleading and needs a comment. In practice, greenhouses are effective because the air trapped inside them is denied the natural process of convection. In mitigation, however, it is the case that the glass panes transmit visible radiation and absorb at wavelengths in the infrared. An alternative, but less appealing, term might be 'radiation trapping effect'.

A common assertion is that carbon dioxide is the principal greenhouse gas while, in fact, it is water vapour that occupies this position. The amount of water vapour present in the atmosphere varies between 1 and 4% (i.e. 10 000 to 40 000 p.p.m. in terms of volume) depending primarily on the temperature. Human activity has little, if any, direct effect on the water vapour content of the atmosphere. Carbon dioxide is the next most important greenhouse gas, and the remaining gases then include methane, nitrous oxide (N_2O), ozone (O_3), and chlorofluorocarbons, or as they are better known CFCs. All of these gases, apart from CFCs, occur naturally. We have more to say about ozone, but in a very different context, shortly. The concentrations of the gases in the present atmosphere are summarized in Table 9.2, although the concentrations should not be taken as directly proportional to their efficiency as greenhouse gases.

Table 9.2 The concentrations of greenhouse gases in the atmosphere.

Greenhouse gas	Concentration[a]
carbon dioxide, CO_2	353 p.p.m.
methane, CH_4	1.7 p.p.m.
nitrous oxide, N_2O	0.3 p.p.m.
ozone, O_3	100 p.p.b.
CFC-12, CCl_2F_2	0.5 p.p.b.

[a] These concentrations refer to volume in the atmosphere and are expressed in either parts per million (p.p.m.) or parts per billion (p.p.b.).

A major concern at present is that human activity, in various forms as discussed earlier in Chapter 6, is significantly increasing the natural concentrations of greenhouse gases in the atmosphere. It is far from easy to predict the overall and accumulating effects of these activities. The whole area, particularly with regards to carbon dioxide, generates a considerable amount of discussion – and publicity. However, amongst all of the uncertainties, it may be permissible to distil two facts. First, it is true that human activity is distorting the natural balance of gases in the atmosphere and, secondly, these gases will *add* to the natural greenhouse effect. The key questions are 'What will be the overall effect?' and 'Are the methods we are using to make predictions of the size of the effect reliable?'. A growing scientific consensus is that there could be a rise in the Earth's mean temperature, i.e. global warming. Whether this is already happening or will become apparent – on an as yet undetermined timescale – remains to be seen.

9.3.2 The influence of ozone

There is another important interaction of solar radiation with the Earth's atmosphere which should be mentioned; it is also vital to life. Look at Figure 9.13.

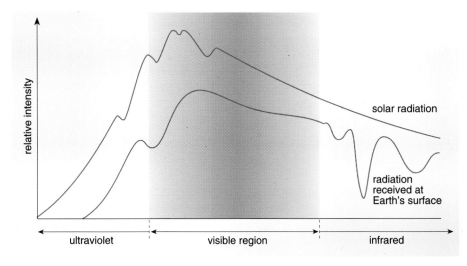

Figure 9.13
The variation of the intensity of incoming solar radiation with wavelength compared to that received at the Earth's surface after passing through the atmosphere.

As expected, the *measured* incoming solar radiation is similar in form to that predicted by assuming that the Sun is an ideal emitter with a temperature equal to its surface temperature (cf. Figure 9.9). Reflection accounts for the general reduction in intensity received at the Earth's surface. The specific absorptions in the infrared region are due to greenhouse gases, with water having a dominant effect in the wavelength region shown. What other major effect do you notice?

You should have noticed that there is a marked cut-off in the ultraviolet region. Outside the atmosphere, the ultraviolet region extends to beyond 200 nm, whereas at the Earth's surface the cut-off is at about 320 nm. In general terms, the ultraviolet region in the electromagnetic spectrum is conveniently categorized into three areas: UV-A between 320 and 400 nm; UV-B between 280 and 320 nm and UV-C below 280 nm. The first is largely

Figure 9.14
Experimentally, the ozone molecule is found to be angular; the bond angle is 117°. In the laboratory, the gas is made by passing an electric discharge through oxygen. It is a blue gas with a pungent smell (the name originates from the Greek word for smell). You may have noticed this smell on occasion when travelling on the London Underground. The 'ozone' at the seaside is more to do with seaweed than ozone gas.

unaffected by the atmosphere and has limited biological effects. The same cannot be said of UV-B and UV-C and it is very fortunate indeed that their intensities are significantly reduced. Both represent highly energetic, short-wavelength radiation. UV-C can interact with, and damage, biologically important molecules in the body. UV-B has various, but not always well understood, effects on humans, animals, plants and materials; it is, for example, implicated in the development of skin cancers.

The reason that most of this harmful radiation doesn't reach the Earth's biosphere is due largely to a single absorbing species: **ozone**, molecular formula O_3 (Figure 9.14).

Ozone is found mainly in the so-called ozone layer which is about 20 km thick and centred on an altitude of roughly 25–30 km, i.e. the stratosphere. Fractional abundances in this layer can reach as high as 10 parts per million. (Ozone is also present at much lower levels in the rest of the atmosphere, cf. Table 9.2, where it acts as a greenhouse gas.) In the stratosphere, ozone is produced naturally by the interaction of ultraviolet radiation of wavelengths below 245 nm with molecular oxygen. This radiation is sufficiently energetic to split up the oxygen molecule into its constituent atoms:

$$O_2(g) \xrightarrow{\text{UV radiation}} O(g) + O(g)$$

The oxygen atoms that are formed are highly reactive and may either react together to reform oxygen, or react with molecular oxygen to form ozone:

$$O_2(g) + O(g) = O_3(g)$$

Ozone itself is a highly effective absorber of radiation in the critical wavelength region of 215 to 295 nm, with some absorptions continuing up to 320 nm. In this way, despite its relatively low concentration, it acts as a very effective filter of harmful UV-radiation. As you are probably aware, there is a considerable concern about the possible depletion of ozone in the ozone layer caused by the release of synthetic chemicals such as CFCs. Since the discovery by the British Antarctic Survey of a massive depletion of ozone – an ozone 'hole' – over the Antarctic continent in 1986, there is concern that ozone depletion could happen in populated mid-latitudes in the future.

9.3.3 Direct uses of solar radiation

As you discovered in Activity 4, an enormous amount of energy reaches the Earth from the Sun. Articles that deal with this subject tend to have their own way of emphasizing this fact. For example, one account notes that the '*total annular solar radiation that falls upon the desert areas of the world (these grow no food and support no population) is estimated to be about 400 times the 1992 world energy consumption of all kinds*'. It would be of great benefit to capture and use some of this energy. A major problem is that technically we are still able to take advantage of only a tiny fraction of what is *directly* available in such abundance.

The simplest, *direct* collecting device for the Sun's rays is a 'flat-plate collector' and this is used to heat water. Such solar panels are often seen on house roofs in sunny countries. The panel is black because black

Figure 9.15
A solar thermo-electric
power station in California

surfaces are efficient absorbers of heat. A more sophisticated approach is to collect the Sun's energy via focusing mirrors. If enough mirrors are used in one locality, then all of the Sun's energy falling on the mirrors can be focused in one place and a furnace created. The furnace can then be used, say, to raise steam for electricity generation (Figure 9.15).

A distinctly more high-technology development is the direct generation of electricity from sunlight via a solar cell which is an example of a photovoltaic device. You may well have one of these in your calculator or watch. These devices are considered in more detail in Book 4. For now, we can simply note that they are able to transform radiant energy directly into electrical energy: no material is consumed in the process and as long as the cell receives light, an electric current flows. Modern solar cells have an efficiency of over 20% in the transformation process, but more than 30% is possible under laboratory conditions. Whole banks of solar cells can be 'added' together to give small power plants mostly under 50 kW, but one plant in California has a peak capacity of around 5 MW and is connected to the grid system. Presently, the cost of electricity from photovoltaic power stations is relatively high compared to that from existing fossil fuel power stations, but this could well change in the future.

And to mention the future is, perhaps ironically, to close this part of Book 2. The development of viable new energy sources is a major concern, but we do not take up discussion of this until later in Book 4. In the next Book, we take a closer look at ourselves and, in particular, our nutrition and health.

> **Question 23** Modelling is very useful in science when dealing with a complicated system. It allows attention to be focused on a key feature of a problem without having to become overburdened with additional detail. Without looking back, briefly explain how a modelling approach emphasizes that the interaction between solar radiation and the Earth's atmosphere must be taken into account when accounting for a mean global surface temperature of 15 °C.

Summary of Chapter 9

Most of the Earth's energy sources depend, in one way or another, on radiant energy provided by the Sun. Non-solar sources of energy include nuclear energy, tidal energy (mainly due to lunar effects), geothermal energy and, conceivably in the future, controlled nuclear fusion. In practice, a distinction is often made between non-renewable energy sources, (e.g. fossil fuels), and renewable energy sources (e.g. those associated with the Sun, the wind and waves).

Electromagnetic radiation is a generic term for all kinds of radiation; they all pass through space at the speed of light and one model views radiation as being wavelike in character. Wavelength (λ) is related to frequency (f) by $\lambda = c/f$ where c represents the (constant) speed of light. Increasing the frequency of radiation shortens the wavelength. Visible light covers the region 700 nm (red light) to 400 nm (violet light). 'White light' is a mixture of all of the visible components. Another model for radiation considers it to be a stream of photons. Photons associated with higher-frequency radiation have more energy associated with them than those of lower-frequency radiation. The radiation emitted by the Sun peaks in the visible region, that from the Earth in the infrared region.

Modelling is useful for demonstrating that the interaction between solar radiation and the Earth's atmosphere is vital in establishing an average and constant global temperature of 15 °C (see Question 23). Many molecules, but not nitrogen and oxygen, will absorb infrared radiation and store it as vibrational energy. This forms the basis of the technique known as infrared spectroscopy.

The greenhouse effect, or radiative trapping, causes an increase in temperature at ground level of some 33 degrees Celsius more than it would be in the absence of the effect. The greenhouse effect is naturally occurring and accounts for a habitable biosphere. The main greenhouse gas is water vapour, with the next most important gas being carbon dioxide. An increase in greenhouse gases due to human activity could result in global warming. Ozone, in the ozone layer, is a very effective absorber of harmful UV-B and UV-C. The gas is formed by the reaction between molecular oxygen and atomic oxygen, the latter being formed as a result of the interaction of ultraviolet radiation with molecular oxygen. There is deep concern that synthetic chemicals are interfering with the ozone layer, causing ozone depletion.

Direct uses of solar radiation are still not particularly well advanced; they include direct collecting devices, focusing mirrors and solar cells which are examples of photovoltaic devices.

Objectives for Part 2 of Book 2

After you have studied Book 2 Part 2, you should be able to do the following:

1   Understand the meaning of the words emboldened in the text.

2   Outline important features of a chemical reaction that are not revealed by its balanced chemical equation.

3   Discuss and give examples of chemical reactions that produce heat when they take place.

4   Discuss the meaning of the term energy and describe how it is stored in various situations.

5   Describe and give examples of energy conservation and transformation.

6   Distinguish between the concepts of temperature and heat.

7   Understand the various units in which energy is measured or quoted and, in addition, be able to interconvert between these units.

8   Recognize the different types of uncertainty associated with scientific measurements and be able to take account of the uncertainties in measured values when carrying out calculations.

9   Outline issues which are associated with the use of fossil fuels as primary energy sources.

10   Describe how chemical processes can be classified in terms of the energy changes that take place when they occur.

11   Understand and use thermochemical equations to describe in a quantitative way the enthalpy changes that accompany either changes of state or combustion reactions.

12   Describe, and illustrate with examples, factors that determine whether a particular fuel is best suited to a particular application.

13   Understand how an enthalpy change for a combustion reaction may be estimated and explain how this procedure provides insight into the exothermic nature of this type of reaction.

14   Describe, in outline form, ideas and concepts that play a central role in discussing why some chemical reactions have a natural tendency to occur and yet others do not.

15   With regard to life in the biosphere, outline and explain in scientific terms key consequences of the interaction of radiation from the Sun with the Earth's atmosphere.

Comments on Activities

Activity 1

Coal, which is a primary energy source, is transported to the power station where it is burned. (Of course, the transport process, itself, involves a whole chain of energy events!) The combustion process produces heat because of the various chemical reactions taking place. So, the first step is the release of chemical energy in the form of heat.

As your own research should have uncovered, this heat is used to raise the temperature of a water boiler so that steam is produced. In turn, the steam powers a turbine which is connected to an electricity generator. Thus, putting aside technical details, chemical energy is converted via mechanical energy (the turbine) into electrical energy. The electricity that is generated then travels through wires until it eventually reaches the electric kettle. Electrical energy is then transformed back into heat in the element of the kettle and this heat finally raises the temperature of the water.

If you did consult a reference source to find out about 'power stations', then it will most probably have mentioned the efficiency of these stations. They are low! An efficiency figure of about 30% (taking into account energy loss in transmission lines and various transformers) is what can be expected. Thus, about one-third of the energy available from the coal eventually reaches your kettle in the form of electricity.

Activity 2

You may have found the following methods:

● Your electricity meter measures the electricity you use in your home in kilowatt-hours. The abbreviation for a kilowatt-hour is kWh. The 'units' on your electricity bill correspond to kWh.

● If you use natural gas in your home, then the quantity consumed is measured by volume – in fact, in hundreds of cubic feet. Various conversion factors are then used so that your gas bill is also expressed in kWh.

● On food labels, 'energy' is expressed in kJ and kcal which stand for kilojoule and kilocalorie, respectively. (You may have noticed that the energy content is usually stated for a given quantity of the food item, often 100 g.)

Thinking more laterally, you may also have suggested we use the 'litre' (or gallon) as a 'measure of energy' when we buy petrol for our cars: indeed, the oil industry uses a 'barrel' as a practical measure of its products!

The more deeply you looked into it, the more you should have come to the conclusion that there is a variety of ways in which energy can be measured in practice.

Activity 3

The idea behind the experiment is to measure – using the electricity meter – the fraction of a kWh that is required to bring an electric kettle filled with water to the boil. However, if the mass of water in the kettle is known, as well as its initial temperature, then the energy required (expressed in J) to raise the temperature to 100 °C can also be calculated. The calculation is based on the fact that 4.184 J is the quantity of energy that is required to raise the temperature of 1 g of water by 1 °C. So, in words,

'energy required' = ('mass of water' × 'temperature rise' × 4.184) J

So, an energy measurement in kWh (via the electricity meter) and in J (by calculation) can be obtained for the same process. Hence, the relationship between the two units can be found.

It is necessary to fill the kettle with a known mass of water. A convenient way of doing this is to take the density to be 1 g cm^{-3} and so pour a known volume into the kettle (1 litre, i.e. 1 000 g, is ideal). The initial temperature of the water in the kettle is also measured. When the electricity is turned on, the rise in temperature of the water is then monitored until 100 °C is reached. There may be some uncertainty in deciding the exact point when this temperature is reached. (In the home, for safety reasons, it is inadvisable to monitor the rise in temperature of the water with a thermometer. An alternative is to rely on 'hearing' the kettle boil which, for the purposes of the experiment, turns out to work quite well.)

For the 'rotating-disc electricity meter', the amount of electricity consumed in the time it takes the kettle to boil is measured by counting the number of revolutions of the disc. However, there is a problem here in that other devices in the home may also be using electricity at the same time. It is necessary to carry out a calibration experiment. A simple approach would be to determine the time for a fixed number of revolutions of the disc, say 10, due to the consumption of background electricity. This would allow you to determine the *average* time for a single revolution of the disc due to the background consumption. If the time taken to boil the kettle in the actual experiment is noted, then the background electricity used in this period can be calculated and so a correction can be made. A good experimental approach would be to ensure that the background consumption was as small as possible.

Some devices – fridges, freezers, central heating pumps – switch on and off intermittently and so the background consumption may not be constant. To avoid uncertainties, these types of device should be switched off for the brief periods of time corresponding to the calibration and actual experiments, but don't forget to turn them back on afterwards.

Activity 4

(a)   There are several ways of approaching the problem, but the strategy must be to compare 'like with like'. One approach is to express world energy consumption in the early 1990s in the units of the watt, i.e. express it in *joules per second*.

Annual world energy consumption is stated as roughly 8 billion (8×10^9) tonnes of oil equivalent. Assuming an average conversion factor of 41.9 GJ (see Figure 4.9), i.e. 41.9×10^9 J, then it follows that

$$\text{annual world energy consumption} = (8 \times 10^9) \times (41.9 \times 10^9) \text{ J}$$
$$= 3 \times 10^{20} \text{ J}$$

This is the energy consumption *per year*. (It is roughly one-tenth of the total energy content of the Middle Eastern oil reserves; see Question 13.)

There are 31 536 000 s in one year ($365 \times 24 \times 60 \times 60$) and this can be rounded to 3×10^7 s. Hence, the world energy consumption *expressed in joules per second*, i.e. watts, becomes

$$\text{world energy consumption} = \frac{3 \times 10^{20} \text{ J}}{3 \times 10^7 \text{ s}} = 1 \times 10^{13} \text{ J s}^{-1}, \text{ or } 1 \times 10^{13} \text{ W}$$

This figure has to be compared with the energy reaching the Earth from the Sun: 1.78×10^{17} W. Solar energy reaching the outline of the Earth was thus some 20 000 times the energy needs of the world in the early 1990s. It is clear that even very inefficient collection of this energy would be worthwhile.

(b)   There are two important factors, along with their implications, to consider. First, the regions in which there is most sunshine (Figure 4.15) – they are mostly desert areas – do not coincide with the areas of high population and industry. And the latter, of course, are the main energy users. Secondly, solar energy is intermittent because of the effects of night/ day and the climate, for example.

The two factors in turn mean that it is necessary to develop – and this involves technology as well as finance – methods of 'collecting' solar energy on a large scale and converting it into forms suitable for long-term storage as well as long-range transmission.

Usually, electricity is a key means of delivering industrial and domestic energy but it doesn't answer all of the problems involved in the use of solar energy. Electricity is difficult to store and transmitting it over long distances is inefficient. One suggested solution to the problems is to produce chemical fuels via solar-driven processes; of these, the production of hydrogen is one possibility. The use of renewable energy sources such as solar energy will be explored further in Book 4.

Activity 5

The general idea of the experiment, as you should have appreciated, is that the heat given out when the methylated spirits burns in air is used to heat the water in the beaker. Furthermore, the amount of methylated spirit that reacts is found by measuring the loss in mass of the simple burner arrangement. Your calculation should have gone along the following lines.

In the experiment, the temperature rise of 250 g of water amounted to 9.0 °C. Given that it takes 4.184 J to raise the temperature of 1 g of water by 1 °C, then the heat released by burning 0.52 g of methylated spirits can be estimated to be

$$(250 \times 4.184 \times 9.0) \text{ J}$$

This gives 9 414 J which when rounded to two significant figures is 9.4×10^3 J, or 9.4 kJ. Since the reaction is exothermic, then $\Delta H = -9.4$ kJ when 0.52 g of methylated spirit is burned in air.

The experiment has quite severe defects:

- some of the heat released by burning the methylated spirits will simply warm up the surrounding air;

- some of the heat released will contribute to warming up the glass beaker in which the water is held;

- not all of the methylated spirits is burned fully to carbon dioxide and water since there is evidence of soot (carbon particles) on the beaker.

In fact, very accurate values of ΔH can be measured using far more sophisticated equipment. A value of the order of $\Delta H = -15$ kJ would be expected when 0.52 g of methylated spirits is fully burned in air.

Activity 6

(a)   The equation for the change of state when ice melts to form liquid water is

$$H_2O(s) = H_2O(l)$$

(b)   The change of state described in (a) is endothermic; it takes energy to melt ice! If experimentally the energy required to melt one mole of water molecules in the form of ice at 0 °C (and normal atmospheric pressure) is 6.01 kJ, then the enthalpy of melting can be expressed as ΔH (melting) = +6.01 kJ. Thus, the thermochemical equation is

$$H_2O(s) = H_2O(l) \qquad \Delta H \text{(melting)} = +6.01 \text{ kJ}$$

(c)   Freezing is the reverse of melting; it is an exothermic change of state. So,

$$H_2O(l) = H_2O(s) \qquad \Delta H \text{(freezing)} = -6.01 \text{ kJ}$$

(d)   The mass of 1.00 mol of H_2O molecules is 18.0 g, i.e.

$$18.0 \text{ g } H_2O = 1.00 \text{ mol } H_2O \text{ molecules}$$

so it follows that

$$1.00 \text{ g } H_2O = \frac{1.00}{18.0} \text{ mol } H_2O \text{ molecules}$$

The energy required to melt 1.00 g of ice at 0 °C must therefore be

$$\frac{1.00}{18.0} \times 6.01 \text{ kJ} = 0.334 \text{ kJ}$$

Hence, for this particular quantity of water, it follows that ΔH (melting) = +0.334 kJ.

(e)   The enthalpy of melting,

$$H_2O(s) = H_2O(l) \qquad \Delta H \text{(melting)} = +6.01 \text{ kJ}$$

is considerably less than that for vaporization

$$H_2O(l) = H_2O(g) \qquad \Delta H \text{(vaporization)} = +40.7 \text{ kJ}$$

The latter change of state involves the water molecules becoming completely separated from one another, whereas in the former the molecules achieve only restricted freedom and so the enthalpy change is consequently smaller.

Activity 7

The steps in the 'recipe' can be followed explicitly.

1   The thermochemical equation for the complete combustion of methane is

$$CH_4(g) + 2O_2(g) = CO_2(g) + 2H_2O(g) \quad \Delta H = -804\,kJ$$

and therefore does involve one mole of fuel molecules.

2   The relative molecular mass of methane, CH_4, is calculated as follows:

relative molecular mass (CH_4) = relative atomic mass (C) + (4 × relative atomic mass (H))

So, relative molecular mass (CH_4) = 16.0

Thus, 1.00 mol of CH_4 molecules has a mass of 16.0 g.

3   According to the thermochemical equation, the enthalpy change is $\Delta H = -804\,kJ$ when 16.0 g of CH_4 is fully burned in oxygen.

4   The enthalpy change when 1.00 g of CH_4 is fully burned in oxygen will be smaller. We know that

$$16.0\,g\ CH_4 = 1.00\,mol\ CH_4\ molecules$$

so it follows that

$$1.00\,g\ CH_4 = \frac{1.00}{16.0}\,mol\ CH_4\ molecules$$

The enthalpy change when 1.00 g of CH_4 is fully burned in oxygen must therefore be

$$\frac{1.00}{16.0} \times (-804)\,kJ = -50.3\,kJ$$

So $\Delta H = -50.3\,kJ$ when 1.00 g of CH_4 is fully burned in oxygen.

5   The heat released when 1.00 g of CH_4 is fully burned in oxygen will be 50.3 kJ.

Activity 8

You should be able to see from the measuring cylinder that the volume of 100 g of unleaded petrol is about 133 cm^3. (This also means the density, i.e. 'mass divided by volume', is about 0.75 $g\,cm^{-3}$). According to Table 6.3, the heat released when one gram of octane is fully burned in oxygen gas is 45 kJ. If this value is taken to be typical for one gram of unleaded petrol, then the calculation goes as follows.

If one gram of unleaded petrol releases 45 kJ of energy, then one hundred grams will release 100×45 kJ = 4.5 MJ. This is the amount of energy released by 133 cm^3 of the fuel. One litre of fuel will thus produce

$$\frac{1\,000}{133} \times 4.5\,MJ = 34\,MJ$$

This can be viewed as the 'energy content' of the fuel. (It is worth recalling that in Box 4.2 it was stated that 'A litre of petrol can provide about 10 kWh'. This is consistent with our calculation because 1 kWh = 3.6 MJ, so that 10 kWh = 36 MJ). Car engines get hot, tyres become warm and you can hear the resistance of the air and wind as a car moves forward. All of these factors, and more, suggest that only a proportion of the energy in the fuel is used to propel the car forwards.

Activity 9

The balanced chemical equation for the combustion of octane gas is

$$C_8H_{18}(g) + 12\tfrac{1}{2}O_2(g) = 8CO_2(g) + 9H_2O(g)$$

At the outset, we can note two features of the equation:

1   The octane is in the gaseous state, which is necessary for the estimation procedure we are going to use. However, in Table 6.3 the octane is in the liquid state and so we must bear this in mind when we come to compare the value of ΔH in the table with our estimate.

2   If we give a molar interpretation to the balanced chemical equation, then one mole of octane molecules reacts with 'twelve-and-a-half moles of oxygen molecules'.

The estimate can be carried out in three distinct steps as described below:

1   The first step is to deal with the bonds that are broken

Bonds broken	Enthalpy change	
$7 \times (C–C)$	7×330 kJ	$= 2\,310$ kJ
$18 \times (C–H)$	18×416 kJ	$= 7\,488$ kJ
$12.5 \times (O=O)$	12.5×498 kJ	$= 6\,225$ kJ
	Total:	$16\,023$ kJ

Note that there are seven C—C single bonds in the octane molecule.

2   In the second step we deal with the bonds that are formed.

Bonds formed	Enthalpy change	
$16 \times (C=O)$ (for CO_2)	$16 \times (-804$ kJ$)$	$= -12\,864$ kJ
$18 \times (O–H)$	$18 \times (-463$ kJ$)$	$= -8\,334$ kJ
	Total:	$-21\,198$ kJ

Already in this step we can see that forming sixteen C=O bonds is going to have a marked influence on the overall *exothermic* nature of the combustion reaction.

3   In the final move, we determine the overall enthalpy change by adding together the 'totals' in 1 and 2 above.

$$\Delta H \text{(reaction)} = 16\,023\,\text{kJ} + (-21\,198\,\text{kJ})$$

$$= 16\,023\,\text{kJ} - 21\,198\,\text{kJ}$$

$$= -5\,175\,\text{kJ}$$

As we indicated to begin with, this value is estimated for octane in the *gaseous* state. The value quoted in Table 6.3, $\Delta H = -5\,120$ kJ, is for octane in the *liquid* state; nevertheless, the two values are quite similar.

Answers to Questions

Question 1 As a starting point, we can write the reaction between propane and oxygen in the air as

$$C_3H_8 + O_2 \text{ /=/ } CO_2 + H_2O$$

For convenience, we shall omit the state symbols until the balancing task is complete. Starting on the left, we see that propane has three carbon atoms and so we must arrange for the same number to be on the right of the equation, i.e.

$$C_3H_8 + O_2 \text{ /=/ } 3CO_2 + H_2O$$

Considering hydrogen next, and using a similar approach, the equation changes to

$$C_3H_8 + O_2 \text{ /=/ } 3CO_2 + 4H_2O$$

Finally, we must consider oxygen and this gives us the balanced chemical equation

$$C_3H_8 + 5O_2 = 3CO_2 + 4H_2O$$

Including the state symbols, and assuming water is produced as a vapour, gives the final equation

$$C_3H_8(g) + 5O_2(g) = 3CO_2(g) + 4H_2O(g)$$

If you are uncertain about any of the steps in this answer, you should return to the more detailed discussion of balancing chemical equations in Part 1 of this Book.

Question 2 The reactants are magnesium metal $Mg(s)$ and carbon dioxide gas $CO_2(g)$. The products are magnesium oxide $MgO(s)$, and solid carbon $C(s)$. The equation to be balanced is therefore

$$Mg(s) + CO_2(g) \text{ /=/ } MgO(s) + C(s)$$

Ensuring that two oxygen atoms appear on both sides of the equation, i.e.

$$Mg(s) + CO_2(g) \text{ /=/ } 2MgO(s) + C(s)$$

results in an imbalance for the magnesium atoms. However, this is easily remedied to give the final balanced equation,

$$2Mg(s) + CO_2(g) = 2MgO(s) + C(s)$$

In this equation, it is the carbon dioxide which is acting as the oxidizer, i.e. the provider of the oxygen for the magnesium fuel.

Question 3 Faraday's experiment illustrates that in a sense the candle flame is *hollow*. The wax of the candle does not burn directly, but is converted into a 'volatile fuel' by the heat of the flame. This fuel exists in the centre of the flame in the region above the wick and inserting a glass tube simply draws some of it away. It will burst into flames if it is heated once more at the other end of the tube.

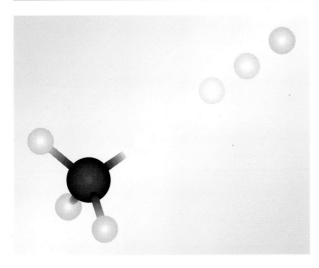

Figure A1
A possible first step.

Question 4 The methane molecule has just four carbon to hydrogen bonds. A reasonable first step, therefore, would be the breaking of one of these bonds as depicted in Figure A1. The result is two highly reactive species: the fragment CH_3 and a free hydrogen atom.

Question 5 The interval which has the largest percentage of oxygen molecules associated with it corresponds to that with the highest horizontal bar: this occurs in the speed interval 380 to $400 \, m \, s^{-1}$ (i.e. *c.* 900 miles per hour). The percentage in the next interval, 400–420 m s^{-1}, is marginally smaller. If we could somehow measure the speed of individual oxygen molecules, there would be the highest probability of finding them in the speed range 380–400 m s^{-1} in the gas at room temperature. This is because, according to the distribution, there is a greater percentage of them in this range. An alternative description is to say that the most probable speeds lie in the range 380 to 400 m s^{-1} rather than in any other 20 m s^{-1} range.

Question 6 The answers to the three questions are also contained in the Summary to Chapter 3. In outline:

(a)   Energy is the capacity to do something. This can be moving or mechanically altering an object in some way. If a temperature difference is involved, then it can be making an object hotter.

(b)   Energy can be stored as kinetic energy or potential energy.

(c)   One property of central importance is that energy can neither be created, nor destroyed; that is the law of conservation of energy. Another is that energy can be transformed from one form into another.

Question 7 You may have made some of the following points or, given the open-ended nature of the question, suggested others.

Basically, human beings rely on plants for food; either we eat them or we eat the animals that feed upon them. The digestion of food essentially involves chemical processes and so changes in chemical energy take place; in fact, into several different forms each vital to life. We need energy for our muscles, in particular to keep us breathing and our hearts beating. We need heat to maintain our body temperature and we need electrical energy to allow our nervous system to function and coordinate. For an individual, there is no doubt: food is vital.

However, in a societal context, a supply of energy is also vital. The processes of modern food production – the manufacture of agrochemicals, the machinery of large scale cultivation, the transport systems for food distribution – all rely on a supply of energy. Without this supply, food production and distribution would be rudimentary and not able to support a society of any size. Only a few would have the ability to be 'self-sufficient'; the rest would starve.

So, the answer is far from clear-cut. Food is essential to life and yet to supply food to a society of any size requires a plentiful energy supply. It would be imprudent to separate either one as being the more important.

Question 8 In powers-of-ten notation,

$1\,000\,J$ can be written as $1 \times 10^3\,J$,

$1\,000\,000\,J$ can be written as $1 \times 10^6\,J$, and

$1\,000\,000\,000\,J$ can be written as $1 \times 10^9\,J$.

If you are uncertain about this notation, you should go back and revise the relevant material in Book 1.

Question 9 Given the definition of the calorie, it follows that $4.184\,J$ (1 cal) is the quantity of energy required to increase the temperature of $1\,g$ of water by $1\,°C$. Given the simplifying assumptions in the question, then the quantity of energy required to raise the temperature of the solution by one degree Celsius can be expressed as $(4.184 \times 100\,J)$. This expression should be treated only as giving a rough value, i.e. about $400\,J$.

Question 10 For an imprecise and inaccurate archery session, the arrows would not be clustered closely together *and* the average position would not correspond to that of the bull's eye. For example, the distribution could be as shown in Figure A2.

Figure A2
An imprecise and inaccurate distribution of arrows for practice shots at the bull's eye

Question 11 The significant figures are as follows:

(a) $10.1\,°C$: there are three significant figures.

(b) $600.5\,J$: both zeros are significant and so there are four significant figures.

(c) $62.0\,kJ$: there are three significant figures for this measurement in kJ.

(d) $0.15\,°C$: ignoring the initial zero, or writing as $1.5 \times 10^{-1}\,°C$, reveals that there are two significant figures.

Question 12 The quantity of energy (59.5 J) and the temperature rise (6.28 °C) are both quoted to three significant figures. Thus, as a 'rule-of-thumb', the result of the calculation should only be given to three significant figures. The numerical result of the calculation is given as 9.474 52. So, in effect, this number should be rounded to the second decimal place; i.e. the 'answer' should be reported as 9.47 J °C^{-1}.

Question 13 To make the estimate, it is best to assemble all of the information to begin with.

● According to Figure 4.8, the proved reserves of oil in the Middle East in 1993 amount to 662.9 thousand million barrels. A thousand million in powers-of-ten notation can be expressed as 1×10^9. Thus, there are 662.9×10^9 barrels (note there are four significant figures).

● According to Figure 4.9, 1 toe is equivalent to about 41.9 GJ, i.e. 41.9×10^9 J. (There are three significant figures here.)

● As an estimate (see Figure 4.9), we can assume 7.3 barrels of oil per tonne. (There are two significant figures here.)

To carry out the calculation, we need to express the Middle East oil reserves in tonnes, that is we divide 662.9×10^9 by 7.3. This number of tonnes multiplied by the energy equivalent (41.9×10^9 J) then provides the answer.

So, in numerical terms we need to find

$$\frac{662.9 \times 10^9}{7.3} \times (41.9 \times 10^9)$$

The answer will be in J and it should contain only two significant figures. So, finally, the estimate becomes 3.8×10^{21} J.

Question 14 The marked differences between the two substances serves to emphasize the influence that hydrogen-bonding has on the properties of water. There will be intermolecular forces between methane molecules in liquid methane but they will be relatively weak. (The most important contribution will be due to London forces.) Hence, liquid methane has a much lower normal boiling temperature and a relatively small enthalpy of vaporization.

Question 15 Steam coming into contact with the (relatively cold) surface of the skin will convert back to the liquid: i.e. condensation occurs. The process is exothermic. The energy released will be over 2 kJ per gram of water condensed. Thus, there is double jeopardy. The condensed water is very hot *and* a significant additional amount of heat is released when it condenses. The overall result is damage to skin tissues.

Question 16 Table 6.2 is shown in completed form below (Table A1).

Table A1 The completed form of Table 6.2.

Initial state	Change	Final state
solid	melting or fusion	liquid
solid	sublimation	gas
liquid	freezing	solid
liquid	vaporization/evaporation	gas
gas	condensation	liquid
gas	deposition	solid

Question 17 To answer the question, it is necessary to ensure that the comparison is made between the same amounts of solid carbon and anthracite. It is convenient to consider kilogram quantities.

For solid carbon, the heat produced by complete combustion of 1 kg will be

$$1\,000 \times (33\,\text{kJ}) = 33\,\text{MJ}$$

Recalling that 1 kWh = 3.6 MJ, it follows that the heat produced – which we interpret as the heat value – for the complete combustion of 1 kg of anthracite will be

$$(9.88 \times 3.6) = 36\,\text{MJ}$$

where the answer is rounded to two significant figures.

The two values are in reasonable agreement with one another. In part this reflects the high carbon content of anthracite. But even so, it should be remembered that this type of coal has a complex chemical structure and to represent it as 'just carbon' is very much a simplification.

Question 18 Your table should be similar to Table A2 below.

Table A2 Hydrogen as a fuel?

Advantages	Disadvantages
On a 'mass-for-mass' basis, a good fuel	On a 'volume-for-volume' basis, it is not as good a fuel as natural gas
Combustion process is 'clean' in pure oxygen	It is a synthetic fuel; energy is needed to produce it
On a small scale, potentially a useful means of storing energy	In contact with air, it forms an explosive mixture if ignited
The explosive nature of the combustion can provide considerable thrust	It is uneconomical to use the refrigerated liquid as a fuel, except in special cases
The gas can be stored safely in metals and alloys	Combustion in air produces nitrogen oxides, NO_x

Question 19 Table 6.3 shows that the enthalpy change per gram of methanol is less than one-half of the typical value for petrol. Thus, unless methanol vehicles are designed to be more efficient it will take twice as much methanol as petrol to travel a given distance. The use of lighter (but stronger) materials in the manufacture of motor vehicles, however, will probably more than compensate for the extra mass of fuel to be carried. As the densities of methanol and petrol are similar, fuel tanks will have to be larger for methanol-powered vehicles.

Other points you may have come up with are:

● Methanol is a synthetic fuel and so the processes used in its production must be economic and environmentally acceptable.

● Methanol is, unfortunately, a 'drinkable' liquid but with tragic consequences because it is acutely toxic.

● Unlike petrol, methanol will absorb water vapour so that fuel tanks made of ordinary steel will corrode.

Question 20 Possible strategies include:

● Further refining to increase the proportion of hydrocarbons with high octane numbers in the petrol mixture: in this case, the process of reforming will be particularly important.

● Blending in octane enhancers. These should be fuels which in their own right have high octane numbers. Oxygenates such as MTBE which are more environmentally acceptable compared to lead compounds, are important enhancers.

● Although not explicitly discussed, you may also have suggested modifying car engines, notably by reducing compression ratios. However, this would not be popular because of the loss of engine power that would result.

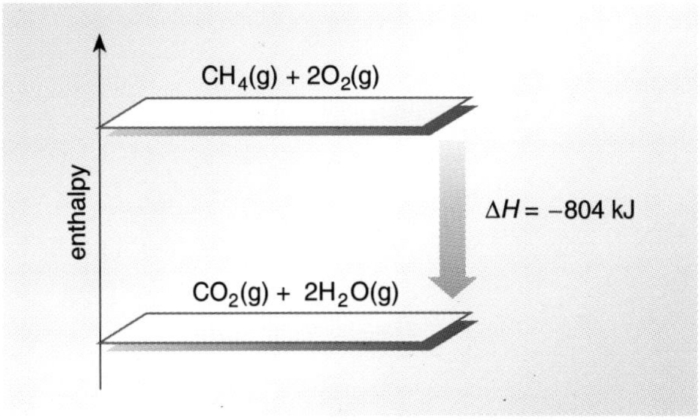

Figure A3
An enthalpy ladder for the complete combustion of methane gas.

Question 21 The complete combustion of methane gas is represented by the thermochemical equation

$$CH_4(g) + 2O_2(g) = CO_2(g) + 2H_2O(g) \qquad \Delta H = -804\,kJ$$

The enthalpy ladder is shown in Figure A3. It emphasizes that the reaction is exothermic since the rung occupied by the reactants, $CH_4(g)$ and $2O_2(g)$, is above that of the products, $CO_2(g)$ and $2H_2O(g)$. The separation between the two rungs is equal to 804 kJ. You should also remember that the ladder represents the reaction as if it occurs at constant temperature.

Question 22 If the blue light and red light have the same intensity, then which one has the higher energy will depend on the energy of the respective photons. Blue light has a shorter wavelength than red light, and so its photon has the greater energy. The answer is blue light.

Question 23 The simplest model to develop is one in which the atmosphere is *not* included. In this model, both the Sun and the Earth are taken to be ideal emitters; the former with a surface temperature of 6 000 °C and the latter emitting radiation in *all* directions back into space. The key point is that there must be a balance between incoming and outgoing radiation in order to maintain a constant average temperature. Given this idea, the model allows an effective temperature for the Earth's surface to be calculated. It turns out to be some 33 degrees Celsius below that of the actual value. It is this failure of the simple model that emphasizes that the interaction between the Earth's atmosphere and solar radiation must be included in a more realistic model. This leads to the idea of the greenhouse effect.

Acknowledgements

Grateful acknowledgement is made to Dr Peter Atkins for providing background material to Chapter 8, and to the following sources for permission to reproduce material in this part of Book 2:

Main cover photo Tony Stone Images/Richard Elliott; *Frontispiece and Figures 1.1 (oil rig), 4.10, 6.8 and 6.16* Science Photo Library; *Figure 1.1 (trees)* Tony Stone Images; *Figure 2.3* British Gas; *Figure 2.4* Allsport; *Figure 2.5* J. A. Conkling (1990) in *Scientific American*, July, copyright © 1990 W. H. Freeman and Co; *Figure 2.9* P. W. Atkins (1991) *Atoms, Electrons and Change*, W. H. Freeman and Co; *Figure 3.4* P. W. Atkins and J. Beran (1994) *General Chemistry* (2nd edn), W. H. Freeman and Co; *Figure 3.8b* IBC Technical Services and W. G. Richards; *Figure 3.8c* Royal Institution and Prof. C. R. A. Catlow; *Figure 3.9b,c Chemistry in Britain*; *Figure 3.13* Royal Society; *Figure 4.5* Hutchinson Library; *Figure 4.9 (lower)* Drake Well Museum; *Figure 4.12* BP Statistical Review, 1994; *Figure 4.14* National Power; *Figure 6.4* Nestlé Products Technical Assistance Co. Ltd.; *Figure 4.15* I. Dostrovsky (1991) in *Scientific American*, December, copyright © 1991 W. H. Freeman and Co.; *Figure 6.11* Hulton Deutsch; *Figure 6.12* NASA; *Figure 6.13* International Academy of Science; *Figure 6.20* Prof. Chris Sheppard, Leeds University, Department of Mechanical Engineering; *Figure 6.24* Sue Cunningham Photographic; *Figure 6.25* Lola Cars Ltd.; *Figures 8.1 and 8.3* ICI Nobel's Explosives Co. Ltd.; *Figure 9.2* Mary Evans Picture Library; *Figure 9.3* JET, Culham; *Figure 9.15* Earth Images.

Index